RANDOM MAPS

RANDOM MAPS

The world in over 100 unusual maps

Simon Kuestenmacher

OH EDITIONS

CONTENTS

PREFACE
The Joy of Maps

Every day, countless people work with maps: city planners, geologists, delivery drivers or members of the military — not to mention cartographers. But there are also many hobbies — I'm thinking of mountaineering or travelling for example — that bring people into deep contact with maps. Finally, the smartphone revolution has turned us all into daily map readers. Where's the next supermarket, what's the fastest way to my destination, will I make it home before it starts raining? These questions can be answered within seconds by an app. We are all map junkies now.

At the same time, there are still many people who are a bit scared of data and maps. Perhaps such fears are rooted in old trauma caused by math and geography lessons in school. But not all data comes to us in the shape of scary mathematical formulas. Sometimes, all it takes is a well-conceived visualisation to make data fun.

My own fascination with maps began as I leafed through my German school atlas. In this venerable tome, complex data was visualised so clearly that it took me on endless mental journeys. Later, during my studies of geography, I constantly worked with maps. Using a computer, I created maps myself. Cartographic software has become much easier and more intuitive since my first lecture in cartography, and more and more data are readily available for free online.

As a consequence, the internet offers a near infinite number of wonderful maps. There is a wide range of forums and blogs where professional cartographers and talented amateurs put their creations on display.

In the last five years I more or less accidentally became the unofficial map curator of Twitter. While doing research for my work as a demographer and journalist in Australia, I constantly stumbled upon exciting maps. They were rarely relevant to my work, but it would have been a shame to keep these remarkable finds to myself. So, in 2016 I started to share a few good maps with my then 100 Twitter followers. I was amazed how many people shared my love for maps. Since I started, about 220,000 new followers have joined the original 100, and every month considerably more than 25 million people view my maps on Twitter. Recently, I've also begun to share the maps on Facebook.

My intention is not to influence my followers politically or force my own opinions onto them. These maps are meant to add to the intellectual framework that you use to form your own judgments. Any map is only an imperfect depiction of an infinitely complex reality. The Polish-American scientist and philosopher Alfred Korzybski warns us never to confuse the map with the territory, just as 'the word is not the thing'. When we confuse maps with territories, we confuse simplified models with reality itself. The same thought has been expressed in an even wittier way by the American philosopher Alan Watts: 'The menu is not the meal.'

Therefore, the purpose of this book is not to pretend that these maps represent reality in a perfectly accurate way. Rather, I hope to awaken in my readers the joy of maps and data. For each of the 100 maps in this book I provide a short explanation of its content and highlights, but I'm happy to let my readers make up their own minds as to their meaning.

Have fun!

POPULATION DENSITY

The three most common types of maps that we encounter in our daily life are topographic maps (showing the shape of the Earth's surface), political maps (showing the borders of countries or areas) and population maps.

Knowing where people live and work and how this is going to develop in the future is of great interest for governments aiming to fairly distribute their limited resources. But companies and marketing agencies are just as keen to know where people will settle down.

It's a well-known adage among cartographers that many maps dealing with all kinds of topics are really just population maps in disguise. The number of mobile phone towers, vehicle licenses, wheelchair users or pets is often just a direct reflection of the population. Obviously, more people means more mobile phone towers and those people will purchase more pets too. Therefore, many such maps are not particularly informative. Clever cartographers get around this with a simple trick: they calculate the number of cats per 1000 inhabitants, or mean cat density, because this number says a lot more about the pet preferences of different areas than the absolute number of cats.

Since population maps are so common and of such interest to many of us, we'll start with some particularly creative examples of this kind of map.

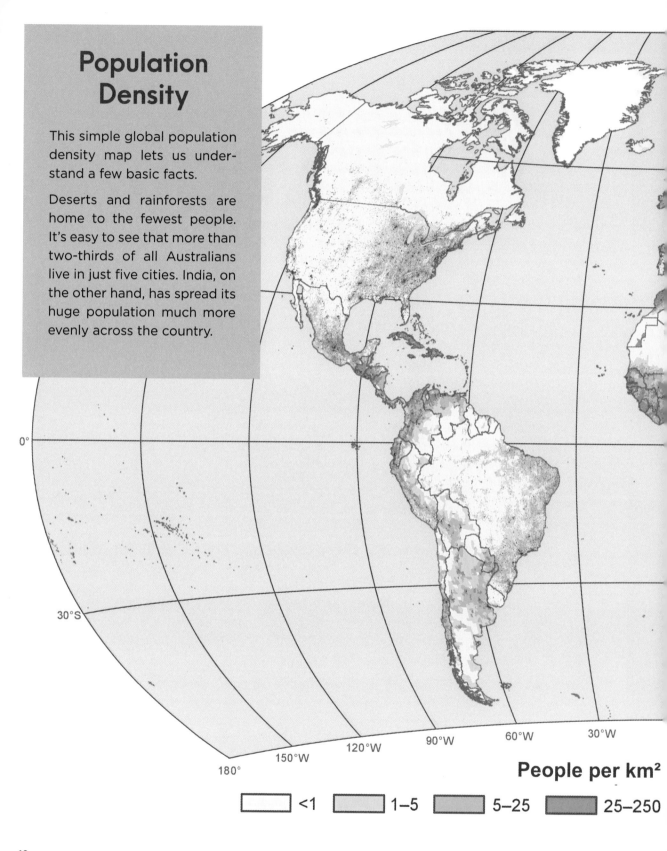

Population Density

This simple global population density map lets us understand a few basic facts.

Deserts and rainforests are home to the fewest people. It's easy to see that more than two-thirds of all Australians live in just five cities. India, on the other hand, has spread its huge population much more evenly across the country.

0°

30°S

180°

150°W

120°W

90°W

60°W

30°W

People per km²

<1   1–5   5–25   25–250

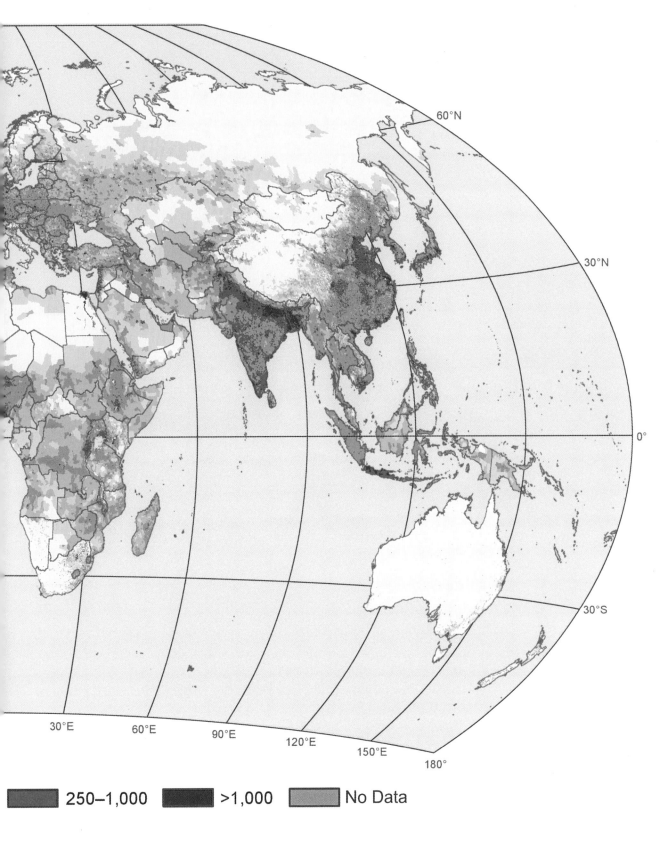

60°N

30°N

0°

30°S

30°E 60°E 90°E 120°E 150°E 180°

250–1,000   >1,000   No Data

The Number of Continents

If you want to create a map that shows how the global population is distributed across individual continents, you must first decide how many continents there are. The answer is not as obvious as you might think. Depending on the definition of the term continent, there are five, six, seven or even eight of them.

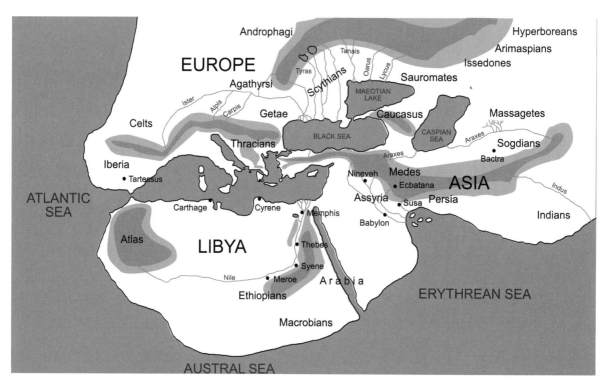

According to the worldview of the ancient Greek geographer Herodotus (490—430 BC) there were only three continents: Europe, Asia and Libya (which we today would call Africa). Strictly speaking, he actually recognised only two continents, because he regarded Libya as a part of Asia. This view is of course outdated since present-day cartographers are aware of the existence of the Americas and Australia.

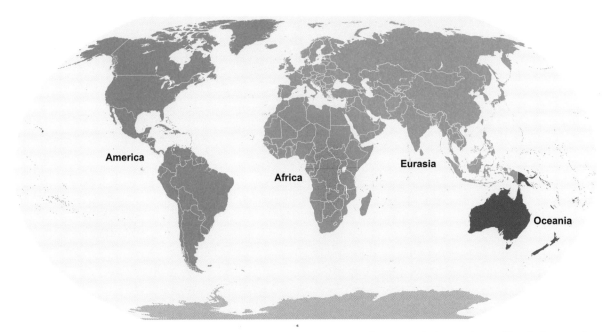

So how about four continents? Do America, Eurasia, Africa and Oceania suffice? From the point of view of plate tectonics, this definition is not without flaws. After all, the continent Eurasia consists of several plates and the Eurasian plate is merely the largest of them. Not to mention that it feels strange for a European to not call Europe a continent . . .

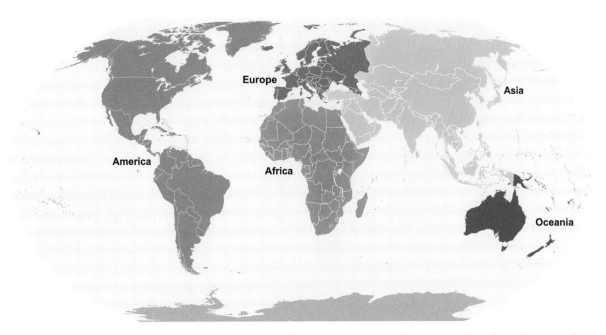

If we consider Europe to be a separate continent, we arrive at the number five. A view that has been popularised by the Olympic rings. The Olympic logo is so pleasing to the eye, can't we just be satisfied with five continents?

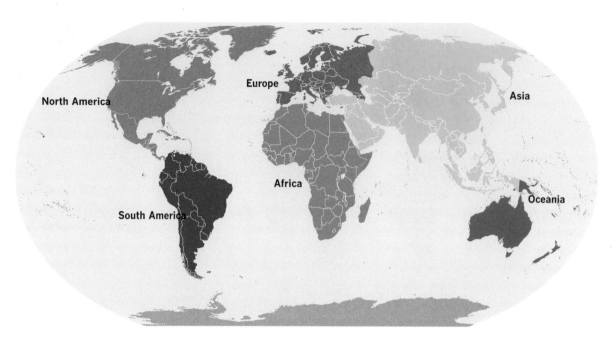

Does It feel fair to count the Americas as a single continent? Isn't South America clearly separated from Central and Northern America? And thus, we have six continents.

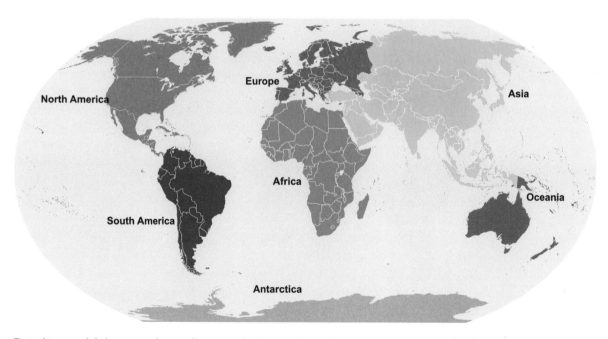

But it would be cruel to disregard Antarctica. So, we get a total of seven continents. There, done!

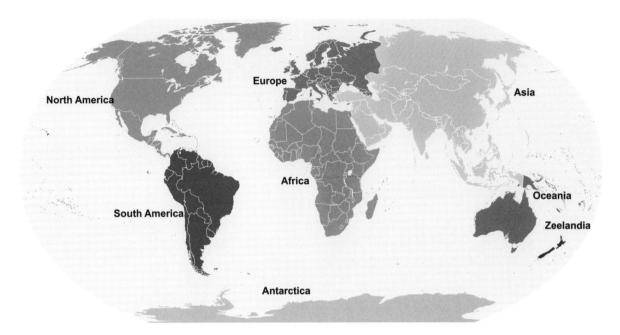

Hold on a minute! Geologically speaking, we have eight continents. The continental crust to which New Zealand belongs satisfies all the scientific criteria to count as a continent, Zealandia. In everyday usage, however, the term continent isn't defined by geological criteria but by a motley combination of cultural, geographical and historical contexts. The example of Turkey shows how difficult it is to distinguish one continent from another. In France and its former colonies, Turkey is generally regarded as a part of Asia, whereas in Germany it tends to be seen as a part of Europe. When it comes to classifying Turkey, there are political as well as tectonic disagreements.

Personally, I prefer the version that counts six continents, for the simple reason that it's the definition accepted by the United Nations. The United Nations Population Division in New York produces the best global population data. Using their definition saves us a lot of time, when trying to depict continental numbers, because it means we don't have to convert data that assumes six continents into data for five or even eight continents. Let's keep it simple.

Population per
Continent in 2020

This map shows us the population per continent (as defined by the UN) in 2020. Western news media often skip over Africa (1.3 billion inhabitants) entirely and covers all of Asia (4.6 billion inhabitants) with glib blanket statements ('China is the new superpower'). Considering how many people live in Africa or Asia, our collective Western ignorance regarding these two continents seems naïve and arrogant. After all, the cultural West (Europe, Northern America and Oceania) is home to a mere 15 per cent of the global population.

Population Growth per
Continent from 2020 to 2040

In its population forecasts the UN states that between 2020 and 2040, Africa will grow by precisely one Europe. Moving every single European to Africa would be a logistic nightmare. The logistical challenges created by the natural population growth within the African continent over the next two decades won't be much smaller though. How can African states supply their citizens with enough electricity, water and education? Europe, even without moving its entire population to Africa, will shrink. How is the European economy supposed to grow with a smaller and aging economy? If we look only at global population numbers, we could assume that the economies in Oceania and North America are going to be the most stable ones in the next two decades. Constant and moderate population growth is a sure sign of an economic boom.

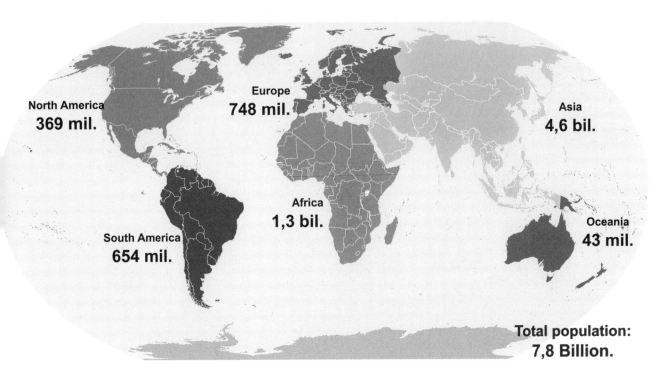

North America
369 mil.

Europe
748 mil.

Asia
4,6 bil.

Africa
1,3 bil.

South America
654 mil.

Oceania
43 mil.

Total population:
7,8 Billion.

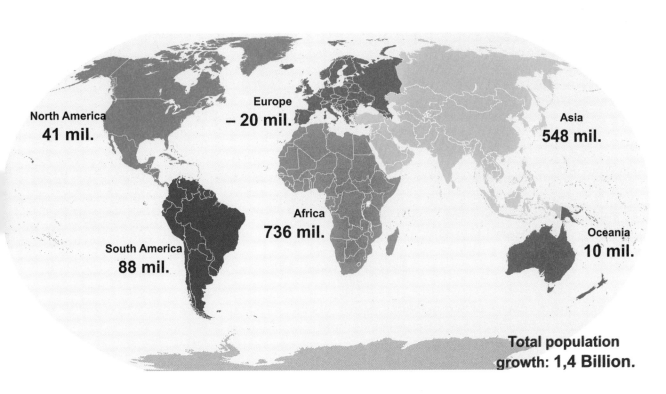

North America
41 mil.

Europe
− 20 mil.

Asia
548 mil.

Africa
736 mil.

South America
88 mil.

Oceania
10 mil.

Total population
growth: 1,4 Billion.

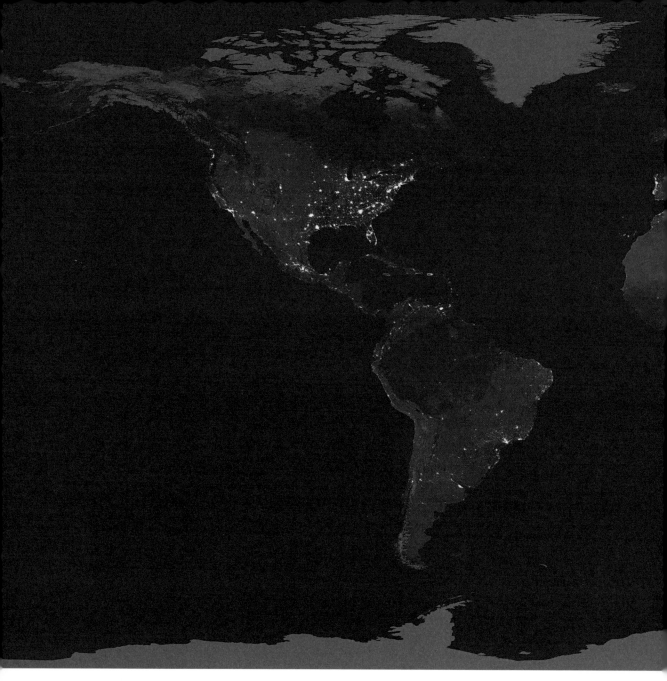

Black Marble

This is how NASA depicts the population density of our planet. 'Black Marble' is a composite image of several satellite pictures taken at night, combined so that clouds don't obscure our view. Global population centres glow yellow, land masses appear dark blue, and the oceans are black. Small Ethiopian villages, for example, are not lit brightly enough to be visible on this map. Therefore, this does not give us a perfect representation of population density. Rather, we can tell which areas are home to the people who can afford to light up the night.

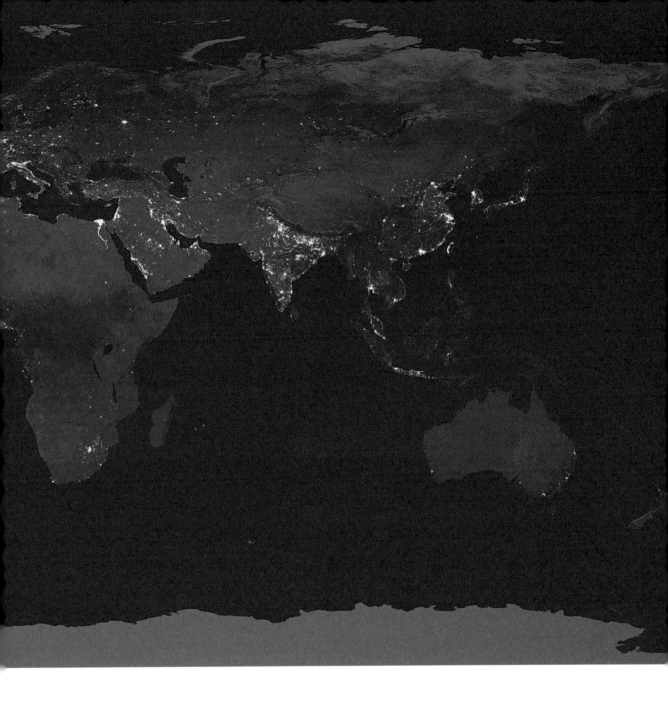

This contrast between rich and poor is very clearly visible on the Korean peninsula. Rich South Korea glows bright yellow, whereas the roughly 25 million people living in poor North Korea spend their nights in darkness. In Egypt, the population lives along the river Nile; the Brazilian and African rainforests on the equator are largely unpopulated and shrouded in the deepest black; Australia is the most sparsely populated continent; and Hawaii and the Canary Islands are little dots of light in the dark ocean.

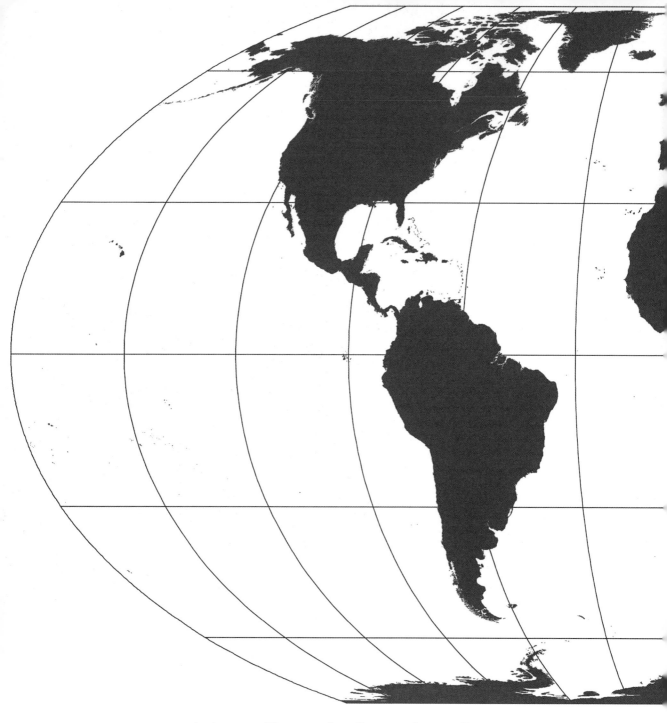

More People Live Inside
the Circle than Outside of It

Humanity is spread unevenly across Earth. More people live inside the blue circle than outside. In other words, it is statistically unlikely for a human to have been born outside this

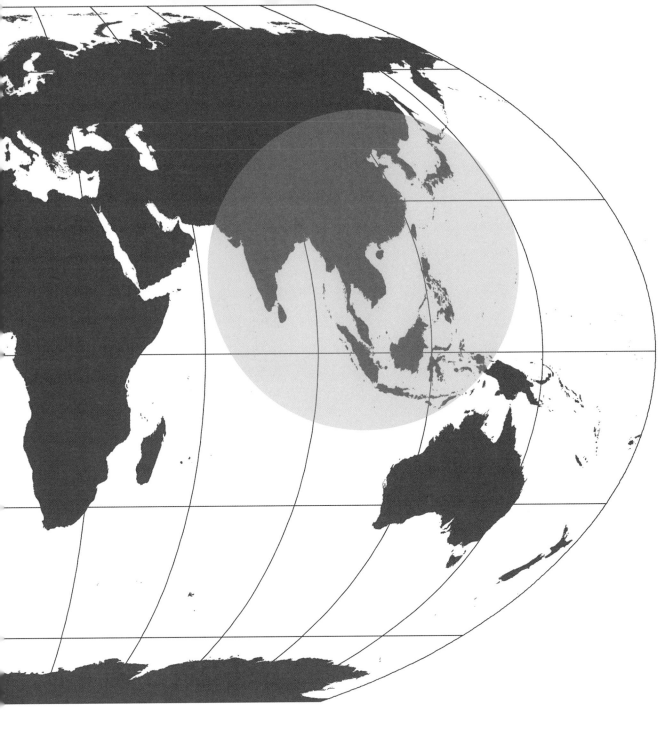

circle. And even within the circle, the distribution is anything but balanced: the circle includes a lot of water, deserts, and highlands. But it is also where we find the megacities of China and India, as well as Indonesia and Japan.

GLOBAL MEGATRENDS

Let's stick with maps that represent the world in its entirety for a moment. It's a good exercise to look at several maps of the same geographical region so we can spot patterns and speculate about possible connections.

Of course, there's no single correct way to view the world. Rather, the point of this exercise is to integrate individual maps and isolated information into one complex world view.

On the following pages we will consider a few selected world maps, which, when viewed together, could significantly change our worldview. These maps look at the world from very different perspectives. A few represent the present, others the past and one of them wildly speculates about the future. How we interpret these maps depends on our expert knowledge, our socialisation and our outlook on the world.

Based solely on the maps shown here: where would you most like to live? Where will the economy remain stable within the next few years, and where will investors risk everything? While these maps can't give definite answers to these questions, they may offer some new perspectives.

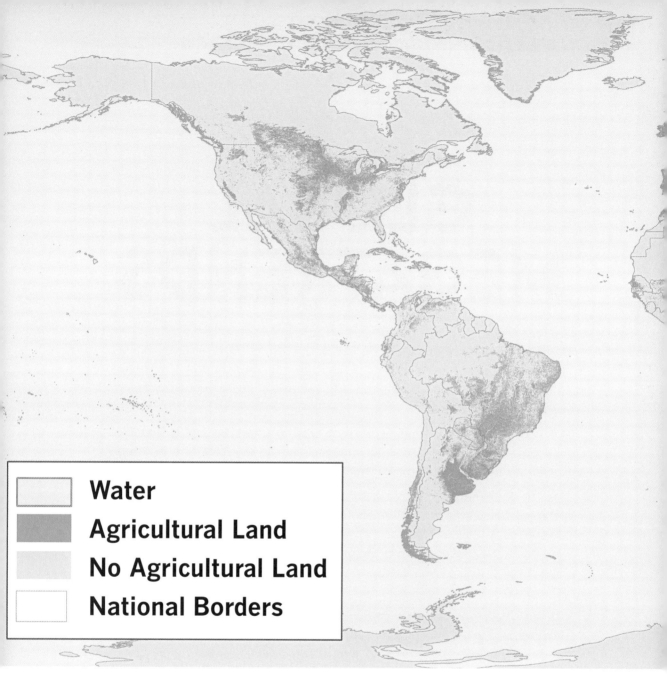

Arable Land

This world map shows the available areas of farmland. In many regions, expanding cities compete with the most arable land. A look at areas of rapid growth such as India, the Chinese east coast or the expanding big cities of Europe reveals this conflict. Humans have expanded the natural boundaries of agriculture in many regions. For example, there are so many artificially irrigated fields in the middle of the Saudi-Arabian desert that we can even spot them on a world map.

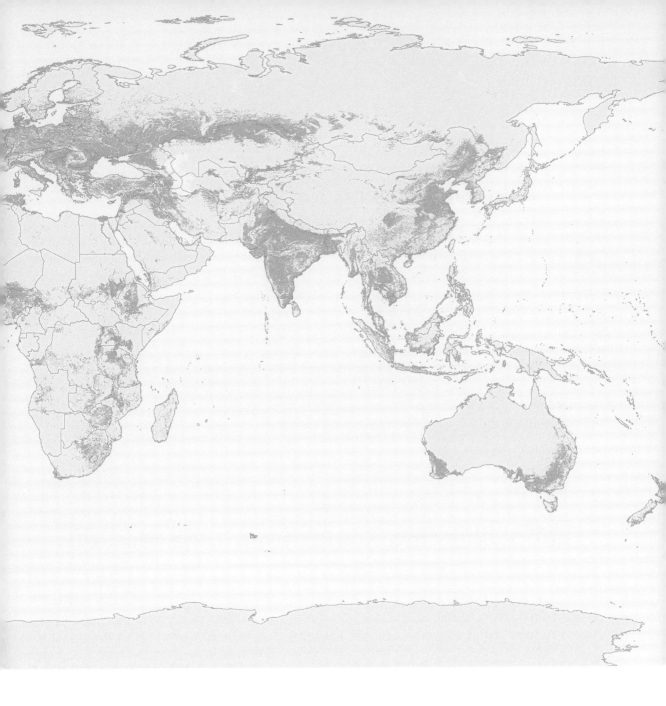

The water that makes this possible comes from a very limited supply of groundwater which will soon run dry. Although this desert nation invests heavily in seawater desalination plants, the amount of water generated in this way is not enough to sustain water-intensive wheat production. Agriculture remains easiest on existing arable land. That is why the protection and restoration of such areas is crucial. In Africa, rising population growth comes up against relatively little arable land. This makes smart land management in the coming decades one of the world's most urgent priorities.

Global Shipping Routes until 1860

These maps visualise the coordinates of historical shipping routes. Ben Schmidt, assistant professor at Northeastern University in Boston, analysed millions of individual coordinates from thousands upon thousands of ship's logs. The first time period shows global shipping routes up to the year 1860. The connections between Europe and the US are clearly visible. Below the southern tip of Africa runs the route along the so-called 'Roaring Forties'. This refers to the zone of westerly winds between the latitudes of 40 and 50 degrees in the Southern Hemisphere. In this zone, powerful winds blow from west to east all year long, causing changeable weather, rain and high seas. There are few safe harbours along the Roaring Forties (Tasmania, New Zealand's South Island, and Patagonia), which made it a very dangerous route. A risk many sailors were willing to take as it shortened travel times significantly, at least as long as you sailed with the wind.

Global Shipping Routes from 1945 to 2000

The second map illustrates how much the world has changed. A bigger, richer and more connected global population engages in increased trade. Shipping has become more reliable as sailing vessels have been replaced by oil-powered ships. Shipping routes have also been reshaped by two of the biggest and most impressive construction projects of the last 200 years: the Suez Canal in Egypt, which opened in 1869, and the Panama Canal, which opened in 1914. Thus, the Roaring Forties have become quiet again, and no one needs to go around Cape Horn anymore.

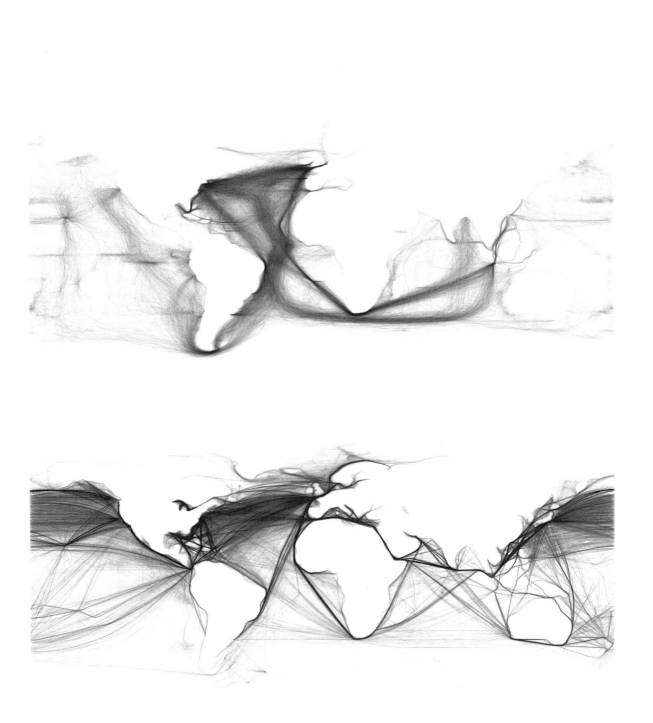

Earthquakes

More than 80 per cent of all earthquakes occur along the, so-called 'Pacific Ring of Fire'. This map shows all earthquakes within the last 100 years around the Pacific. Its geographic location forever influences a country. For example, due to their location on the Pacific Rim, New Zealanders have to consider completely different architectural and infrastructural risk factors than their Australian neighbours.

NORTH AMERICA

SOUTH AMERICA

PACIFIC OCEAN

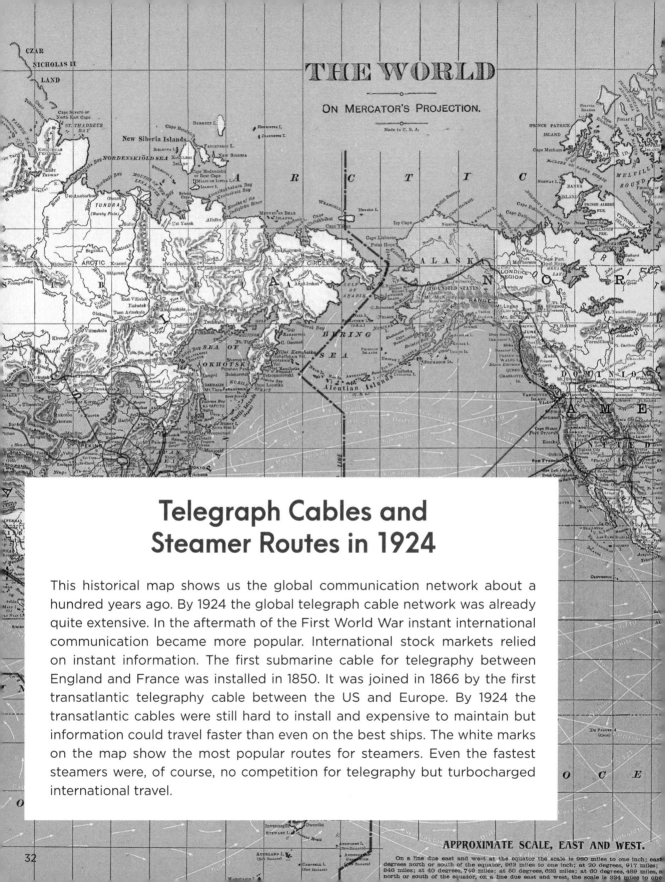

Telegraph Cables and Steamer Routes in 1924

This historical map shows us the global communication network about a hundred years ago. By 1924 the global telegraph cable network was already quite extensive. In the aftermath of the First World War instant international communication became more popular. International stock markets relied on instant information. The first submarine cable for telegraphy between England and France was installed in 1850. It was joined in 1866 by the first transatlantic telegraphy cable between the US and Europe. By 1924 the transatlantic cables were still hard to install and expensive to maintain but information could travel faster than even on the best ships. The white marks on the map show the most popular routes for steamers. Even the fastest steamers were, of course, no competition for telegraphy but turbocharged international travel.

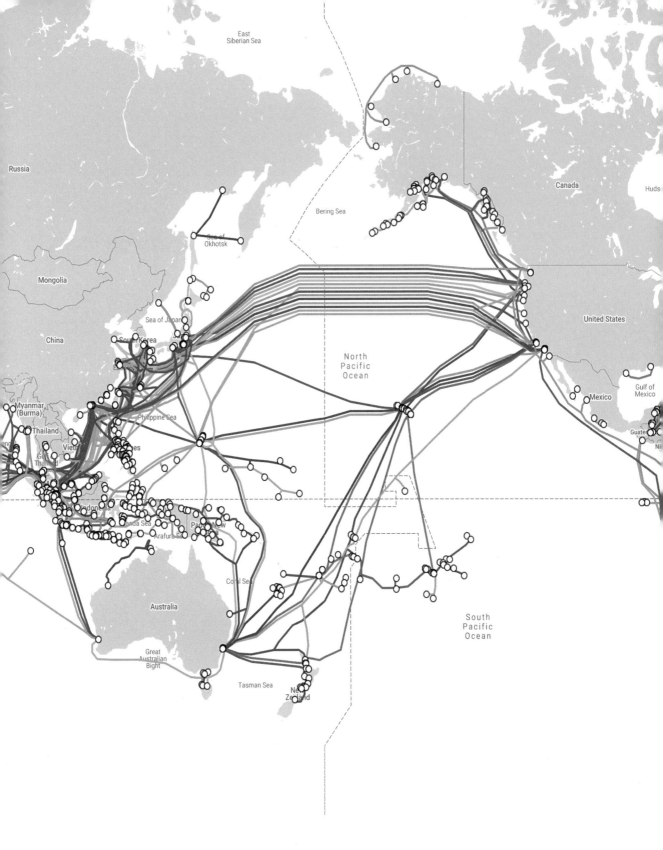

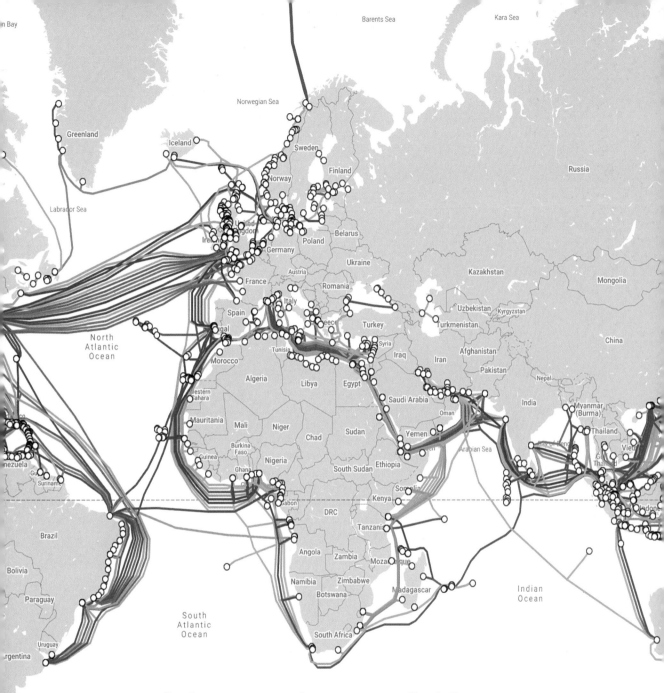

Submarine Internet Cables

Today our global communications network is even faster and more convenient than it was in 1924. A lot of online traffic is only made possible by submarine cables. The volume of data traffic between two economic zones reflects the importance of their economic relations. There is particularly lively data traffic between the North American East Coast and Europe; and the North American West Coast and China and Japan.

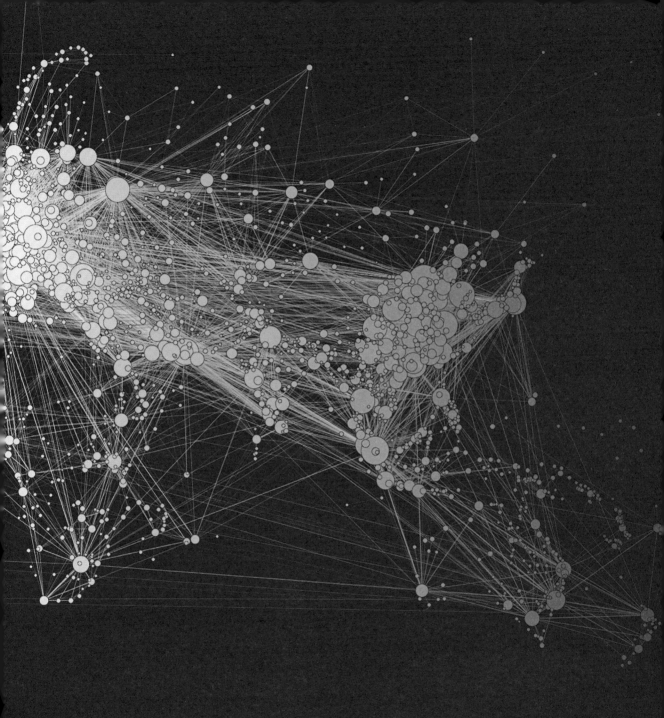

Global Air Traffic

This map shows global flight connections. The Swiss cartographer Martin Grandjean visualises the number of flight connections per airport with the size of the circles and the flight paths themselves with thin lines. The colour of the circles depends on the longitude of the airport and serves a purely aesthetic function.

Arctic Passage
Without sea ice, the shipping route is open all year round, providing transport links between habitable areas in Canada and Russia

Greenland
Greenland's ice sheet is melting rapidly

Scandinavia / Great Northern Russia / G
Dense cities consisting of h provide a home for the maj

Canada
Reliable rainfall and higher temperatures offer ideal growing conditions for most of the world's subsistence economies

Southwest USA
Due to desertification, the last inhabitants of this region are migrating north. The Colorado River is just a small trickle. The area is used for solar energy and geothermal energy

North Africa / Middle East / South of the USA
A solar energy belt stretches for thousands of kilometers using a mixture of photovoltaics and solar thermal energy.
A high-voltage DC power plant sends electricity north at regular intervals

Amazon
Desert

Africa
Mainly desert, although some models show the green of the Sahel

Peru
Dry and uninhabitable area due to de-glaciation

Patagonia
Melting glaciers reveal a new cultivable zone, although the poor soils have to be reclaimed

West Antarctica
Unrecognizable. Densely populated with high-rise cities

© Connectography (2016) by Parag Khanna

Food cultivation / dense cities consisting of high-rise buildings

Uninhabitable desert area

Uninhabitable due to floods, drought or extreme weather conditions

Afforestation potential

Loss of lan assuming

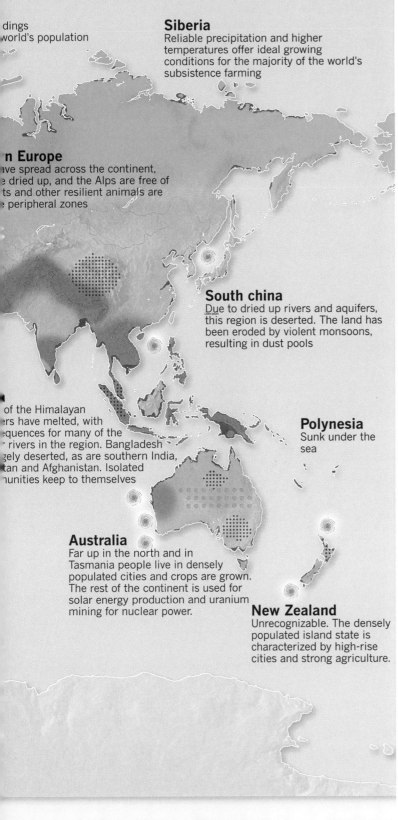

Siberia
Reliable precipitation and higher temperatures offer ideal growing conditions for the majority of the world's subsistence farming

dings
world's population

n Europe
ve spread across the continent,
e dried up, and the Alps are free of
ts and other resilient animals are
e peripheral zones

South china
Due to dried up rivers and aquifers, this region is deserted. The land has been eroded by violent monsoons, resulting in dust pools

of the Himalayan
rs have melted, with
quences for many of the
rivers in the region. Bangladesh
ely deserted, as are southern India,
an and Afghanistan. Isolated
unities keep to themselves

Polynesia
Sunk under the sea

Australia
Far up in the north and in Tasmania people live in densely populated cities and crops are grown. The rest of the continent is used for solar energy production and uranium mining for nuclear power.

New Zealand
Unrecognizable. The densely populated island state is characterized by high-rise cities and strong agriculture.

sing sea levels,
o meters

 Solar energy

 Geothermal energy

 Wind energy

Global Consequences of Climate Change

Geopolitical strategy advisor Parag Khanna uses this map to show the consequences of extreme global warming of more than four degrees Celsius. This isn't the most likely of warming scenarios; many experts suggest an increase of two or three degrees. Extreme climate change doesn't have negative impacts for every single corner of the planet. Some local investors might well benefit from climate change. For example, if you wish to grow wheat in Siberia, the thawing of the ground and increase in fertile land would be grounds for celebration. And less ice in the Arctic is a welcome shortcut for shipping.

However, globally the negative consequences easily outweigh the positive ones. Desertification will make agriculture in southern Europe, much of the US and sub-Saharan Africa almost impossible. This puts the nutrition of millions of people into peril and increases the risk of gigantic streams of refugees and enormous human suffering.

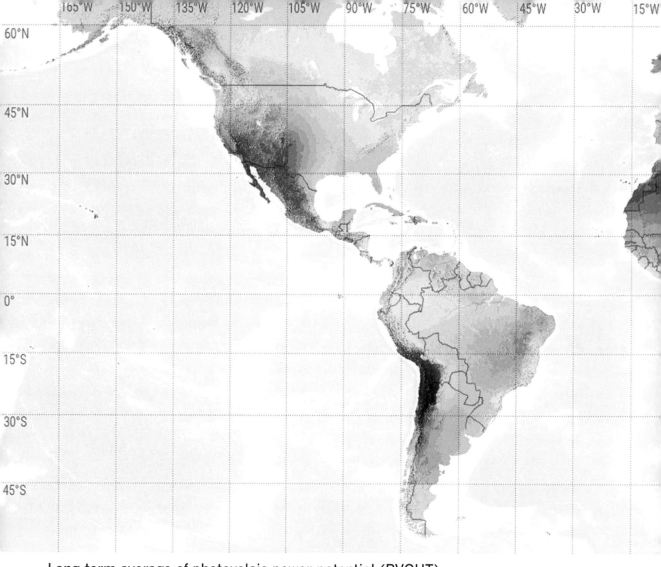

Long-term average of photovolaic power potential (PVOUT)

Daily totals:	2.0	2.4	2.8	3.2	3.6	4.0
Yearly totals:	736	876	1022	1168	1314	1461

Solar Power

We can still counter desertification and climate change. One option would be to switch from fossil fuels to renewable energy, in order to lower our collective greenhouse gas emissions. Solar power can help us generate our energy more sustainably. The map shows how much photovoltaic energy could be generated annually. If you have ever driven through any villages

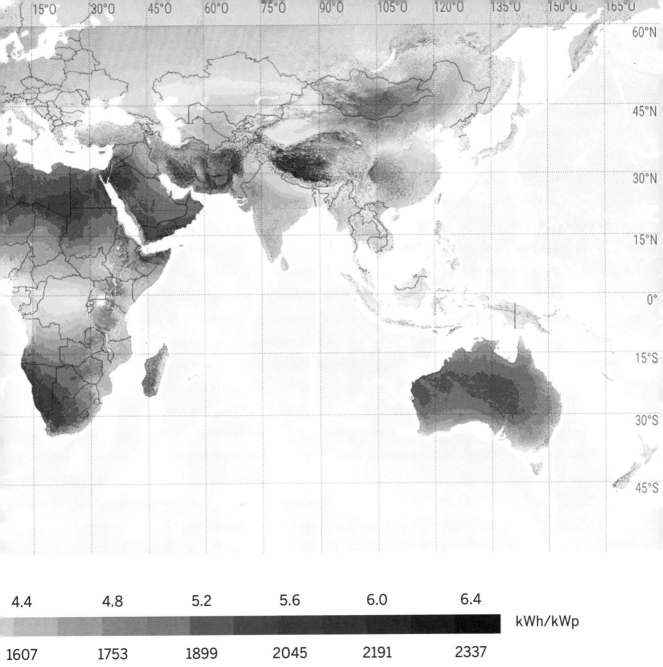

4.4	4.8	5.2	5.6	6.0	6.4	
						kWh/kWp
1607	1753	1899	2045	2191	2337	

in southern Germany (light green on the map) and seen the countless solar panels on the roofs of houses and barns, you will be shocked when you drive through Australia or South America and see barely any of them. These countries could make much more effective use of solar thermal power or photovoltaic systems.

WANDERLUST

When I think back to my youth and school days, I no longer know which came first: my fascination for maps, or my love of travel. These two passions definitely created a kind of positive feedback loop. After leafing through an atlas for a while, I just wanted to venture out and explore.

As with travel, maps are often about exploring and crossing borders in some form or another. Maps tease us with the possibility of coming up against our own inner limits and the opportunities for personal growth. This tantalizing possibility is what makes maps so attractive. While a map lets us dream about the world and makes us long for adventure, it is no substitute for travel. That said, through maps even the biggest couch potatoes can cross national borders, penetrate the unknown and gain new experiences.

In this chapter, we'll explore the world by plane, by car, by bike, as hitchhikers, and even as time travellers alongside Charles Darwin. We won't neglect the practical considerations of a real trip, such as socket adapters and speed limits either.

Visa-free access to . . . countries

	160–189
	130–159
	100–129
	70–99
	40–69
	25–39

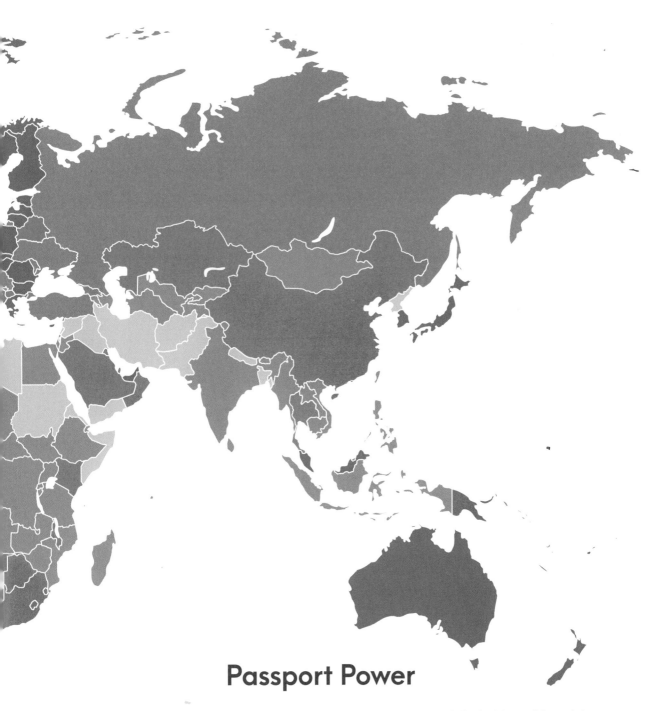

Passport Power

Those who travel usually need a passport. Henley & Partners, a global citizenship advisory firm, has counted how many countries a country's citizens have visa-free access to. In other words: how much do other countries trust the citizens of that country? A Japanese or Singaporean passport grants visa-free access to 189 countries. German and Finnish citizens get access to 187 countries. At the other end of the spectrum are undemocratic, mostly poor countries that are embroiled in conflict. Afghans have access to only 25 countries. Iraqis, Syrians, Pakistanis and Somalis have similarly limited freedom of travel.

Tipping

Travellers know this problem only too well. You've just ordered your first meal in a foreign country and suddenly realise that you have no idea whether it is appropriate to tip and, if so, how much you should give. In Japan and Australia, for example, people don't tip, but the reactions when you do are very different. Australians are happy to accept any additional dollar. In Japan, on the other hand, this is a cultural gaffe as it is considered shameful for the waiter to accept tips.

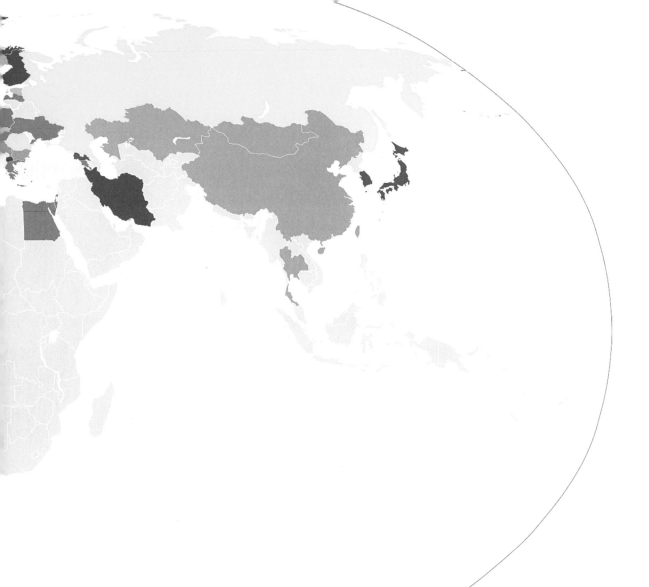

Expectation:	NOT EXPECTED, reaction if tipped:				EXPECTED, reaction if not tipped:	
Attitude:	Insulted	Confused	Neutral	Grateful	Punished	Punished
No tips at all						
Rounding up, <5 %						
5—10 %						
~10 %						
10—15 %						
15—20 %						

Type A   Type B   Type C   Type D   Type E   Typ

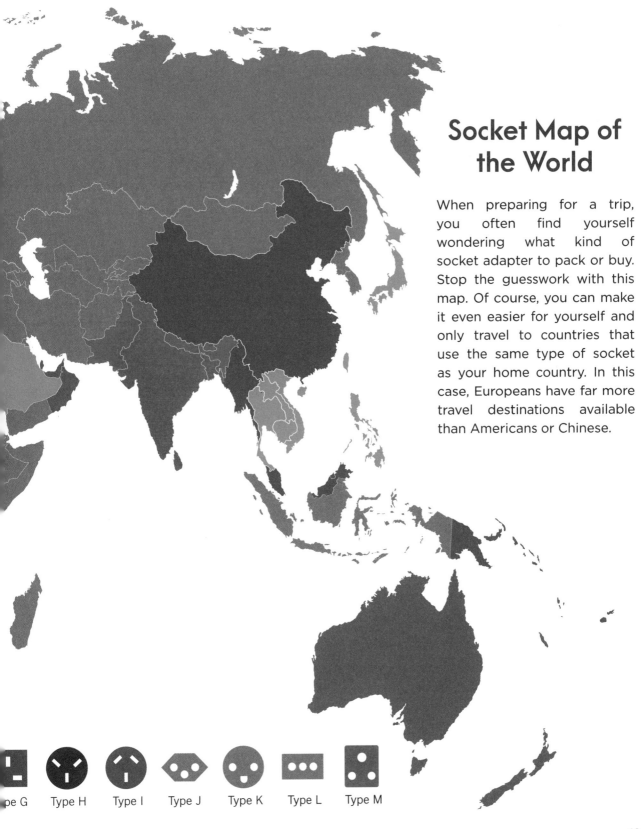

Socket Map of the World

When preparing for a trip, you often find yourself wondering what kind of socket adapter to pack or buy. Stop the guesswork with this map. Of course, you can make it even easier for yourself and only travel to countries that use the same type of socket as your home country. In this case, Europeans have far more travel destinations available than Americans or Chinese.

pe G   Type H   Type I   Type J   Type K   Type L   Type M

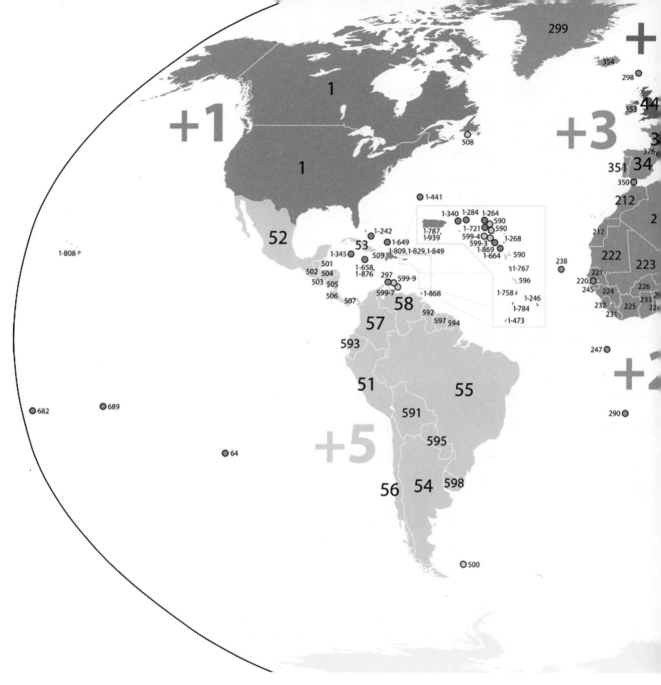

International Country Codes

A cursory look at this map of telephone country codes makes you wonder why the codes are not sorted a little more logically. The reason lies in the history of how the codes were assigned. Initially, phone calls were routed by human operators and international dialling codes were unnecessary; the operator would just make the connection manually. As more and more people called abroad, each country needed its own number. The allocation of the first country codes was regulated in 1960 by the CCITT (Comité Consultatif International Téléphonique

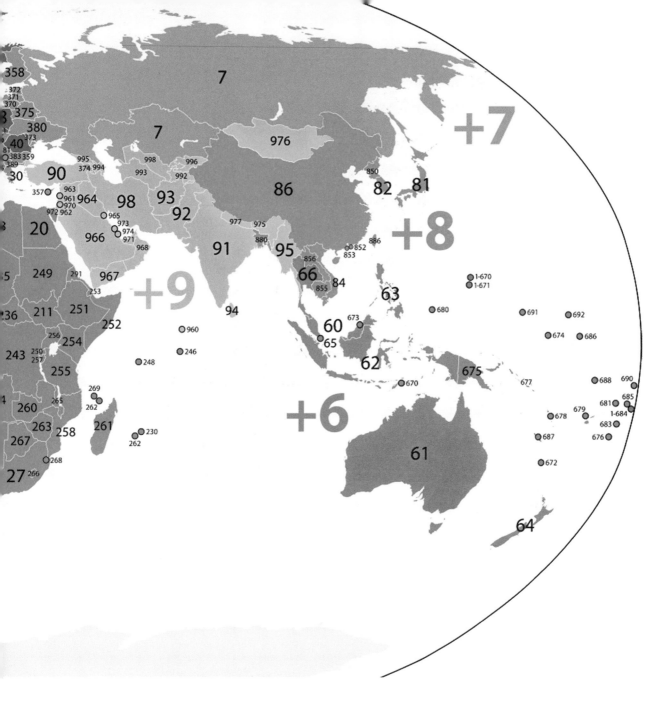

et Télégraphique) in Switzerland. The area codes began with 20 for Poland, through to 49 for the Federal Republic of Germany, to number 79 for the Soviet Union. Numbers were only available for European countries, including the Soviet Union. Once computers were able to make telephone connections automatically, a global dialling system was needed. This happened in 1964. The world was then divided into nine zones and numbered from 1 to 9: 1 for North America, 2 for Africa, 3 and 4 for Europe, 5 for South America, 6 for the South Pacific and Oceania, 7 for present-day Russia, 8 for East Asia and 9 for the rest of Asia.

Left- and Right-Hand Traffic

Before the age of cars, when horses dominated traffic, left-hand driving was almost universal. Mounting from the left was easier as this was the side of the belt that you carried your sword. In Russia, wagon drivers sat on the left rear horse, so they could easily reach all animals with

Driving on the Right
Driving on the Left
No data

their right arm. If you wanted to overtake other wagon, it was safer to do so if everyone drove on the right side. This way the drivers could more easily estimate the distance between the passing wagons. After the French Revolution, right-hand traffic spread across most of Europe. Only Great Britain remained loyal to the left side of the road and through its colonies kept left-hand traffic a worldwide phenomenon.

Literal Meanings of European Country Names

If you have a hard time picking your next holiday destination, just choose a country with an entertaining name. Unfortunately, most countries in Europe are simply named after ancient peoples. 'Land of the Franks'? Sorry France, but I'm looking for something a little more special. I would rather go to Spain, the 'Land of Many Rabbits'. Travellers with poetic sensibilities, on the other hand, are better off journeying to the 'Navel of the Moon' (Mexico), the 'Land beside the Silver River' (Argentina) or 'I Go to the Beach' (Nauru).

ICE LAND
(Iceland)

LAND OF THE SCOTS
(Scotland)

THE LAND
OF ERIU
(Ireland)

LA
OF
AN
(Eng

LAND OF THE
FOREIGNERS
(Wales)

SHRU

PORT OF CALE
(Portugal)

LAND OF MANY
RABBITS
(Spain)

LAND OF THE
SVEAR
(Sweden)

LAND
(Finnland)

NORTHERN
WAY
(Norway)

WATERSIDE
DWELLERS
(Estonia)

FOREST
CLEARER
(Latvia)

FLAT
DERLAND
enmark)

SHORELAND
(Lithuania)

WHITE
RUSSIAN
(Belarus)

OWLANDS
etherlands)

LAND OF
THE PEOPLE
(Germany)

PEOPLE OF
THE FIELDS
(Poland)

LAND ON THE EDGE
(Ukraine)

LITTLE CASTLE
(Luxembourg)

LAND
OF THE CZECHS
(Czech Republic)

LAND OF THE SLOVAKI
(Slovakia)

DARK RIVER
(Moldova)

D OF
GAE

BRIGHT STONE
(Liechtenstein)

EASTERN
REALM
(Austria)

COUNTRY OF
MAGYARS
(Hungary)

PEOPLE
FROM ROME
(Romania)

F
NKS

LAND OF THE SWISS
(Switzerland)

LAND OF PEOPLE
WHO SPEAK THE
SAME LANGUAGE
(Slowenien)

SINGLE
HOUSE
(Monaco)

LAND OF
SAINT MARINUS
(San Marino)

ON THE MOUNTAIN RIDGE
(Croatia)

RIVER BORN
PROPERTY
OF A DUKE
(Bosnia and
Herzegovina)

LAND OF
THE MEN
(Serbia)

HOME OF THE
MIXED TRIBES
(Bulgaria)

PAPAL PALACE
ON THE HILL
(VATICAN CITY)

LAND OF THE BLACK
MOUNTAIN
(MONTENEGRO)

FIELD OF THE
BLACKBIRDS
(Kosovo)

LAND OF
TALL PEOPLE
(Macedonia)

ISLAND OF
COPPER
(Cyprus)

LAND OF YOUNG
CATTLE
(ITALY)

LAND OF EAGLES
(Albania)

PLACE OF
REFUGE
(Malta)

LAND OF THE
HELLAS
(Greece)

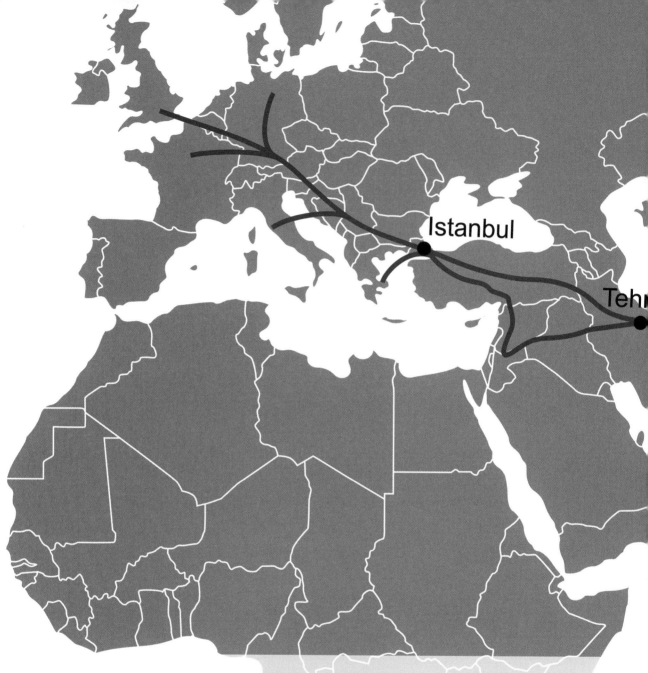

Hippie Trail

In the 1960s young soul-searching travellers headed overland from Europe to Asia, establishing the Hippie Trail. These hippie tourists usually had little money, so hitchhiking was very popular. The Hippie Trail is also generally considered to have started the backpacking trend. The main stops on the route were sites of spiritual growth. Local culture and sights were just as important as the availability of intoxicants. Istanbul was the gateway to Asia and gave the hippies their first glimpse of the exotic world awaiting them on their travels.

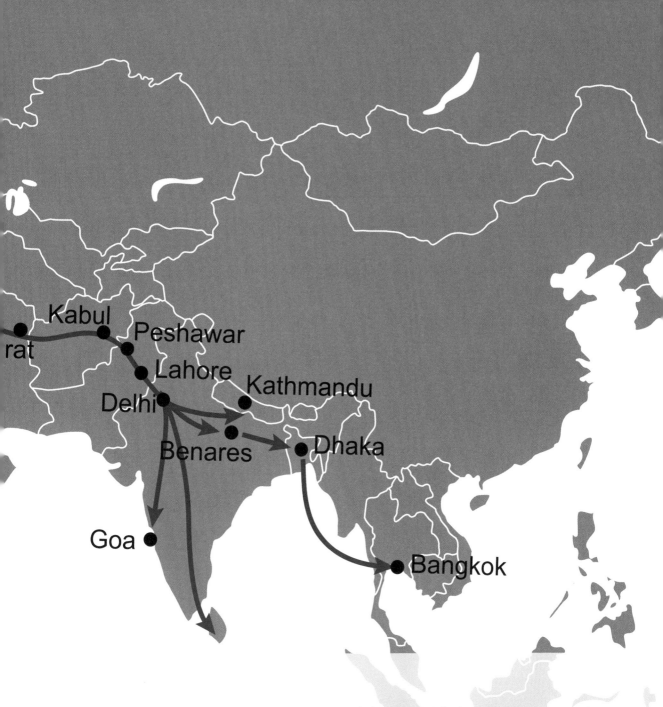

Afghanistan offered cheap hashish, opium, and heroin; Kabul was then a modern, cosmopolitan city. Drugs, temples, and gurus were also available in Kathmandu. The spiritual climax, however, were the ashrams of Goa, where most journeys culminated. Famous hippie meeting places formed at the most important junctions of the trail, where travellers exchanged information about the best transport routes, cheap accommodation, and which natural remedies were worth a try. When the Soviet Union invaded Afghanistan in 1979, the land route between Europe and Asia was cut off and the Hippie Trail became history.

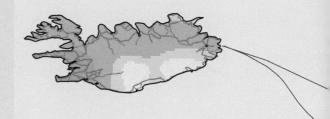

Average Waiting Time for Hitchhikers

In the 1970s, hitchhiking was a common practice worldwide. Despite cheap airlines and Uber, hitchhiking is still a thing though. On this map we can see how long hitchhikers typically have to wait before they are picked up. The best places to hitchhike are countries like Ireland, Romania, Albania, Algeria or Tunisia. But hitchhiking also remains a particularly comfortable way to travel in some more rural regions of Germany. For example, travel from Ravensburg to Villingen-Schwenningen is a breeze. Also, who would've thought that travelling between Manchester and Liverpool was harder than travel in and around London?

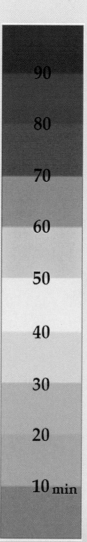

European Long-Distance Cycling Routes

This map shows the official European long-distance cycling routes registered by the European Cyclists' Federation. The routes connect local bike paths to form a continuous network. How about a north-south crossing of Europe from the northernmost point of the continent (the North Cape in Norway) to the southern tip of Italy or even to Malta? That's a whopping 4500 miles (7400 km) of cycling fun. If you don't have three months of leave available in one go, your vacation plans for the next few years are sorted. For amateur historians, I recommend route number 13: it leads you along the former Iron Curtain, which separated Eastern and Western Europe for almost half a century, passing through 20 countries in total — the ideal way to learn about European history of the 20th century.

●Lisbo

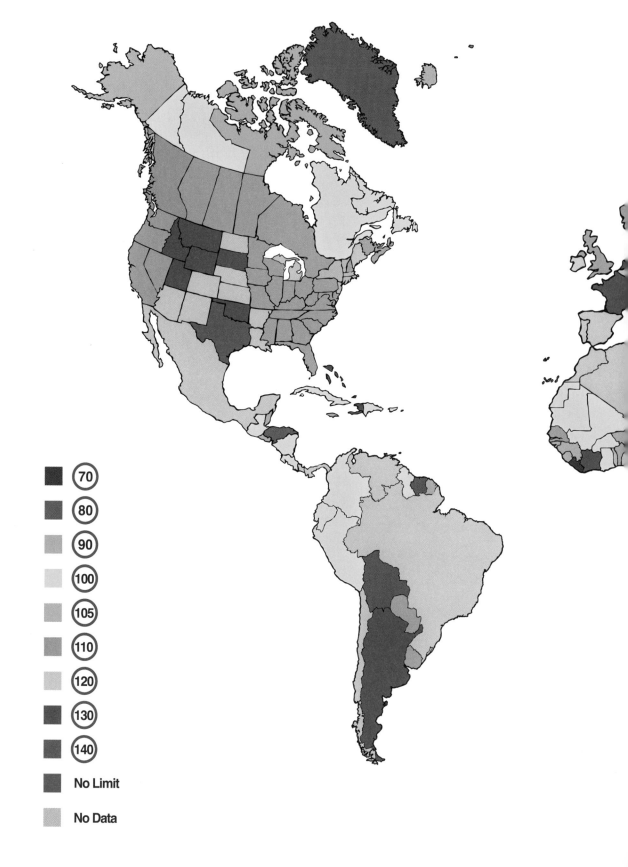

70

80

90

100

105

110

120

130

140

No Limit

No Data

Speed Limits
Worldwide

This map is of interest to travellers who are going by car because it shows the maximum speed limits across the globe. There is only one country without a speed limit — you know who we are talking about here. . . However, many visitors to Germany are quite disappointed when they find out that only a few short sections of the German autobahns have no speed limit. Even the Germans must adhere to speed limits most of the time.

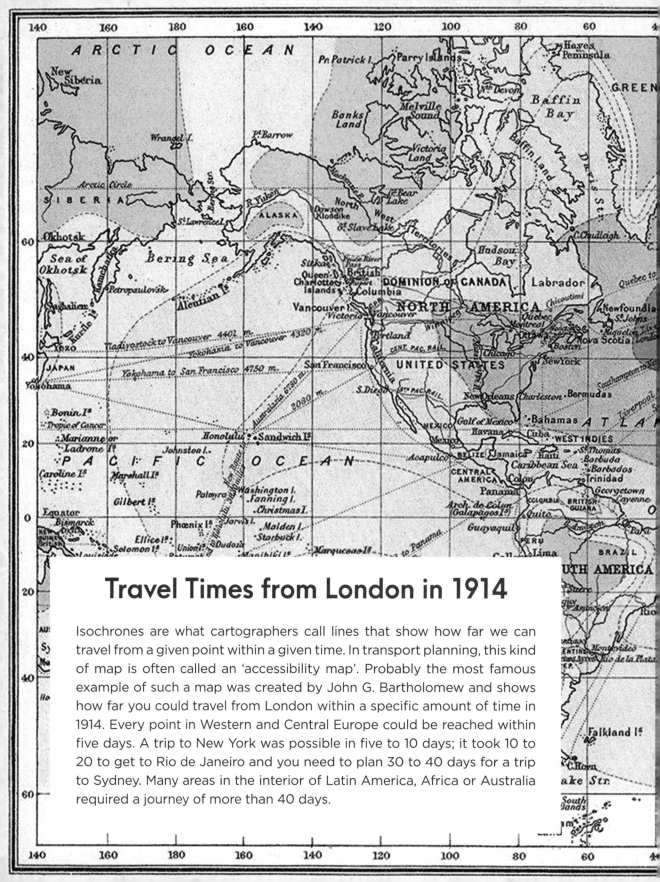

Travel Times from London in 1914

Isochrones are what cartographers call lines that show how far we can travel from a given point within a given time. In transport planning, this kind of map is often called an 'accessibility map'. Probably the most famous example of such a map was created by John G. Bartholomew and shows how far you could travel from London within a specific amount of time in 1914. Every point in Western and Central Europe could be reached within five days. A trip to New York was possible in five to 10 days; it took 10 to 20 to get to Rio de Janeiro and you need to plan 30 to 40 days for a trip to Sydney. Many areas in the interior of Latin America, Africa or Australia required a journey of more than 40 days.

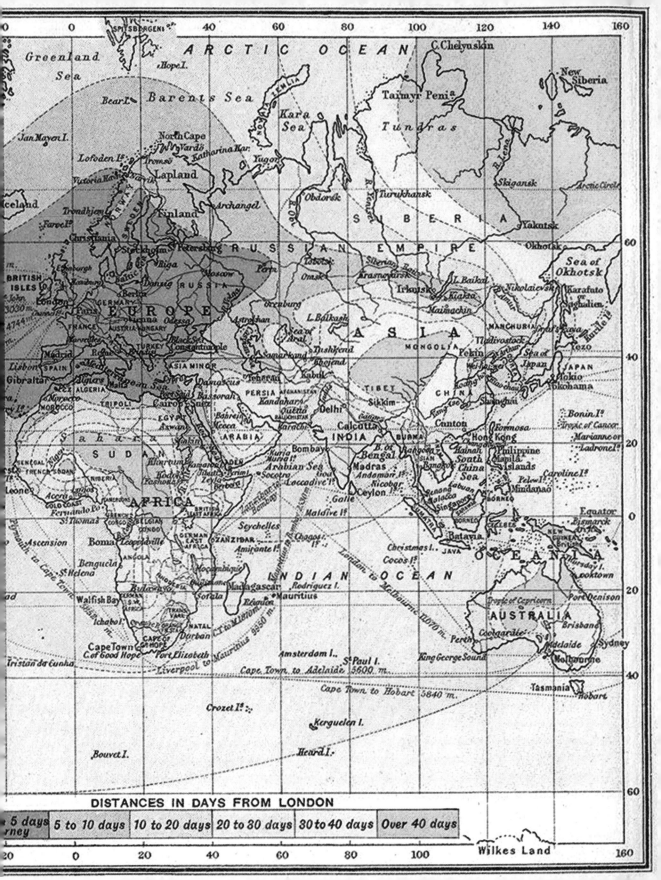

DISTANCES IN DAYS FROM LONDON

| 5 days | 5 to 10 days | 10 to 20 days | 20 to 30 days | 30 to 40 days | Over 40 days |

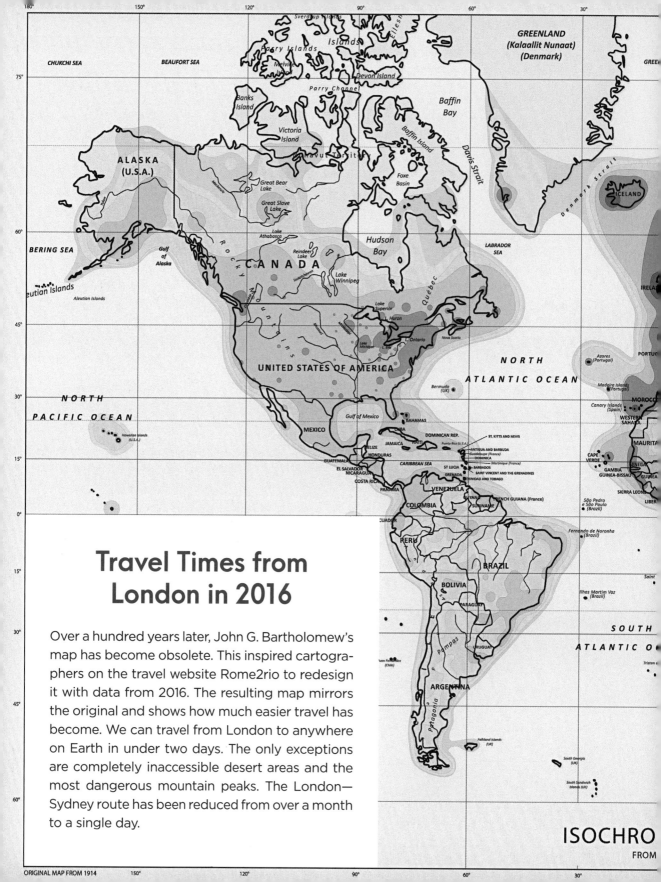

Travel Times from London in 2016

Over a hundred years later, John G. Bartholomew's map has become obsolete. This inspired cartographers on the travel website Rome2rio to redesign it with data from 2016. The resulting map mirrors the original and shows how much easier travel has become. We can travel from London to anywhere on Earth in under two days. The only exceptions are completely inaccessible desert areas and the most dangerous mountain peaks. The London—Sydney route has been reduced from over a month to a single day.

ISOCHRO

FROM

DISTANCE IN HOURS FROM LONDON

| 0-6hr | 6-12hr | 12-18hr | 18-24hr | 24-30hr | 30-36hr | 36+ hr |

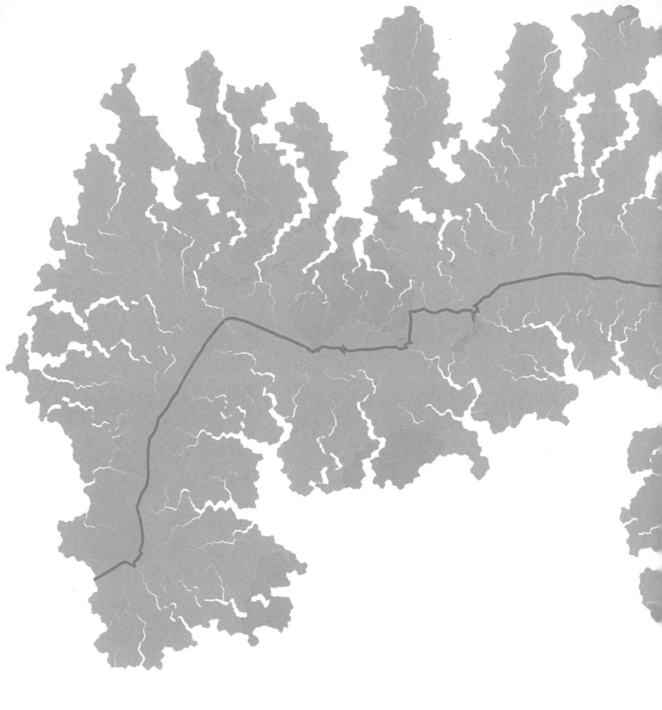

The Voyage of the Beagle

If you have a data set and want to display it on a world map, you need to consider which map projection is most suitable. Every two-dimensional map is a distorted representation of our three-dimensional world. A cartographer must decide which of the known standard projections best fits the data they want to present.

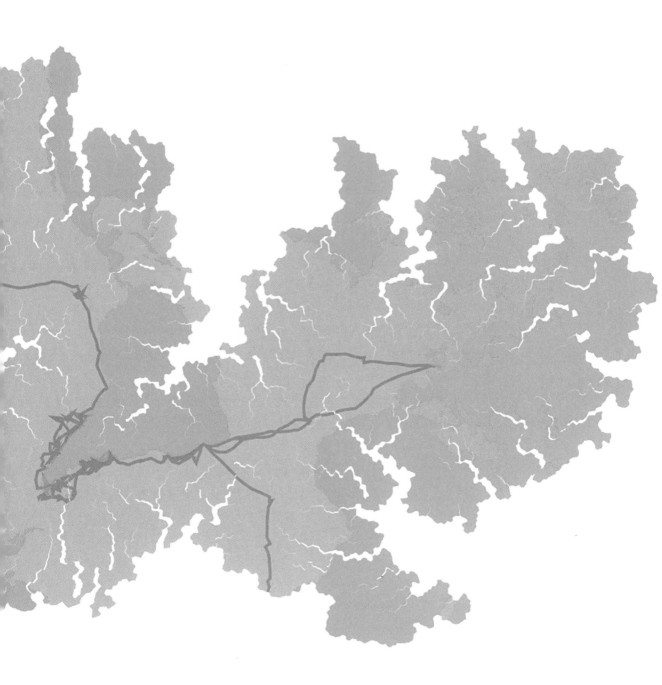

Thanks to modern cartography software, we can even create a unique map projection for a specific set of data. On this map, the cartographer Ben Schmidt shows us the route that Charles Darwin took on the HMS Beagle (1831—36), the voyage which became the basis for his theory of evolution. The map projection has been programmed in such a way that the regions close to the ship's route remain undistorted, while areas far away from the route are shown with high distortion. Finally, the map is oriented so that the Galapagos Islands, which were of particular importance for Darwin's theory of evolution, are in the centre.

Eruption of the Yellowstone Supervolcano

A visit to the world's oldest national park is a must for nature lovers travelling the US. The park was founded in 1872 right on top of a volcano. And not just any little volcano, but a super-sized American volcano. An eruption would result in a huge cloud of dust. The map shows how this dust cloud would develop (under normal weather conditions in April). The dust cloud could extend over 1900 miles (3000 km) to the East Coast of the USA and deposit a layer of dust up to 3 mm thick. Within a radius of about 125 miles (200 km), the layer of dust would be more than a metre thick. However, this would happen slowly over the course of weeks, and geologists don't think that the supervolcano will erupt any time soon. No need to cancel your Yellowstone visit just yet . . .

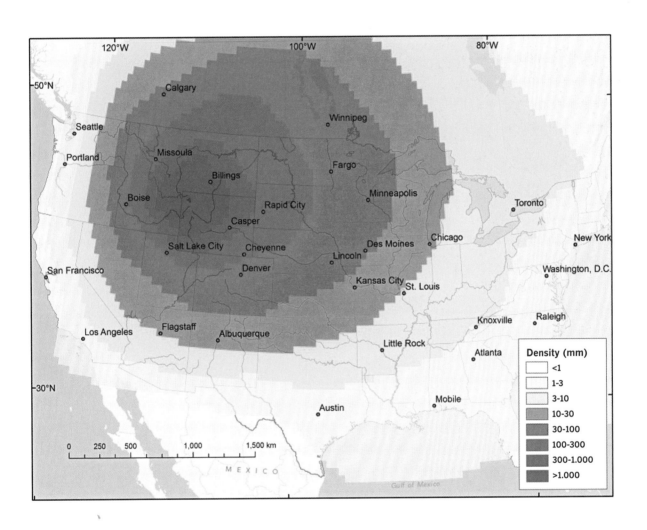

Density (mm)
	<1
	1-3
	3-10
	10-30
	30-100
	100-300
	300-1.000
	>1.000

ANIMALS

The real travel fanatics of this planet are not us humans, but the other animals. Anyone following my Twitter account knows how much I love animal maps. As a little boy, I always wondered about all the things birds get to see in their lives. Today, thanks to GPS trackers, researchers know exactly how animals move — and their data leads to exciting maps.

Cat owners can buy one of these GPS trackers for the cost of a pair of jeans. The latest models weigh less than 100 grams and can be clipped to a collar. A map on their mobile phone shows the owner what secret lives their cats lead outside of the family home.

In this chapter you will discover, among other things: how wolf packs define their territories, where the woolly mammoth lived, which country is home to the most venomous animal species and how dolphins would draw a world map . . . if they could.

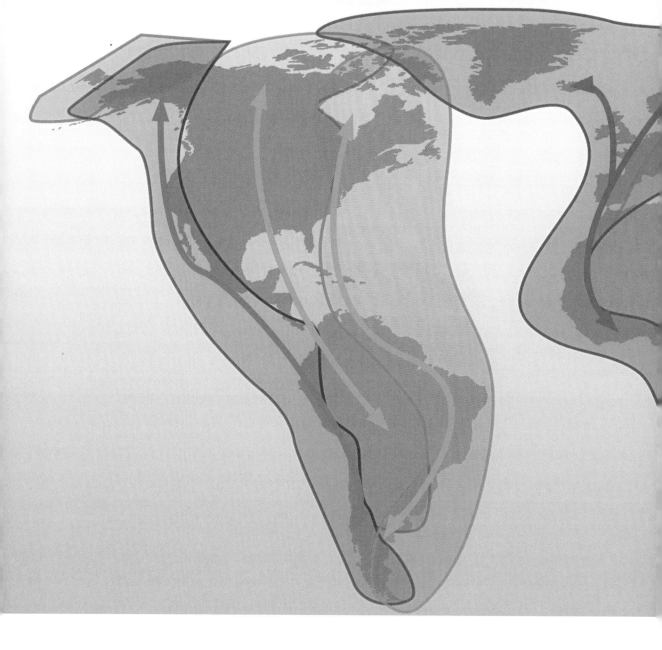

Arctic Migratory Birds

This map shows the eight main migration routes for migratory birds. It is difficult to regulate conservation efforts because you have to consider economic and ecological aspects as well as the interests of numerous different parties. Coordinating all of them becomes even trickier when more than one country is involved. Of course, birds don't care about borders as they cross national airspaces and make stopovers in different countries. People and institutions must be at their best to protect migratory birds. Successful international species protection shows us what we are capable of.

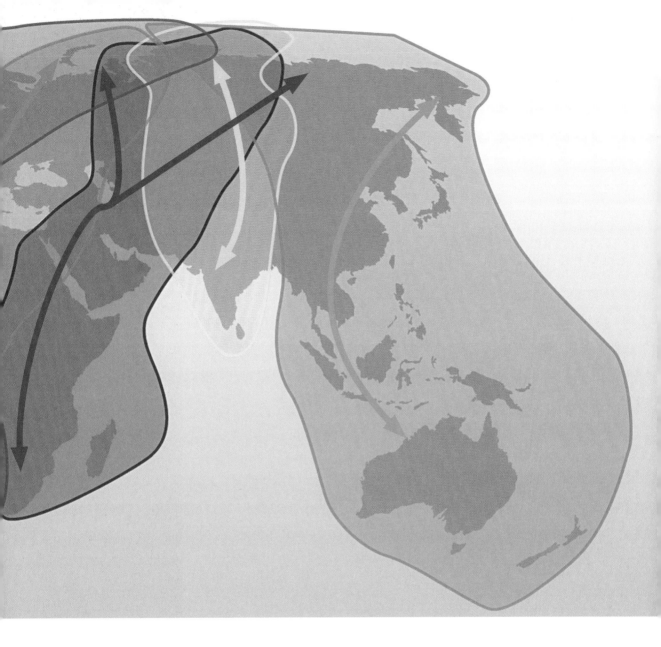

Scientists had no proof that birds migrated until 1822. Without GPS tracking, people could not be sure what exactly happened to migratory birds such as storks in winter. Other theories, besides long distance migration, included the idea that migratory birds hibernate or even turn into other animals such as mice!

Only when several so-called 'Pfeilstörche' (arrow storks) were found in Germany could the migration of birds to faraway places such as Equatorial Africa be confirmed. Arrow storks are exactly what they sound like: storks that survived being struck by the arrow of an African hunter and that still carried that arrow around. The first known case was the Rostock arrow stork, which was brought down in 1822, with an African arrow still stuck in its throat. It was stuffed and can still be viewed at the Rostock Zoological Collection.

Wolf Pack Territories

At first glance this map looks like messy scribbles in Microsoft Paint, but it really displays very interesting data from the Voyageurs Wolf Project. Conservationists fitted wild wolves with GPS-collars. Each coloured scribble represents a different wolf in Voyageurs National Park in the US state of Minnesota on the US-Canadian border. In addition, every GPS-tracked wolf belongs to a different pack. The map clearly shows that each pack respects their neighbours' hunting grounds. The individual territories are each around 70 km² in size. The data on this map tracks the wolves' movements throughout summer, from April to October. During this time, there is a large supply of fawns, hares, beavers and other prey animals and the wolves usually hunt alone. In winter, they band together and hunt large animals such as elk and deer in packs.

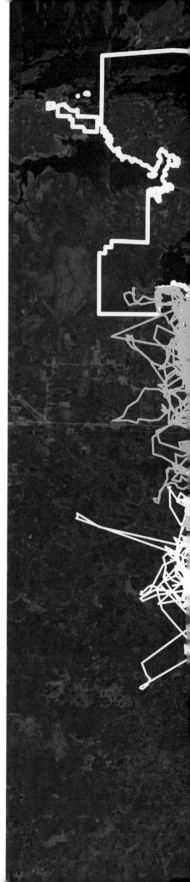

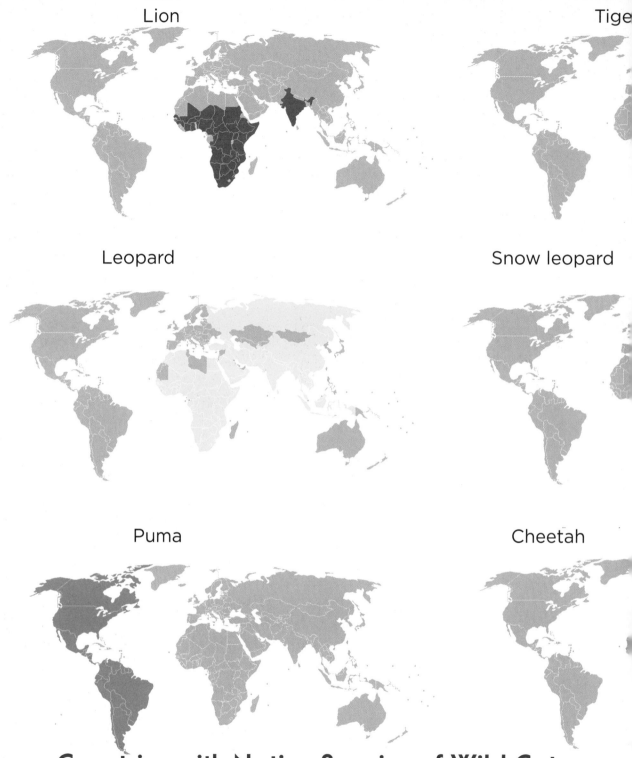

Lion

Tige[r]

Leopard

Snow leopard

Puma

Cheetah

Countries with Native Species of Wild Cats

This series of world maps shows which big wild cats are native to which countries. However, it doesn't depict the actual size of the areas where the big cats live. These would be much

Jaguar

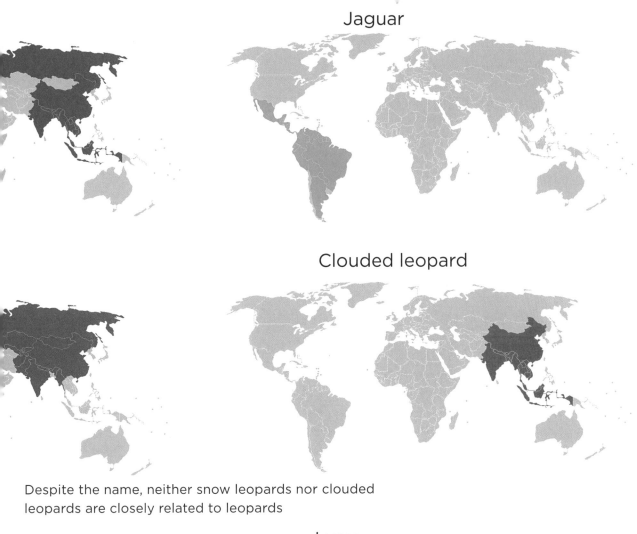

Clouded leopard

Despite the name, neither snow leopards nor clouded leopards are closely related to leopards

Lynx

too small to be clearly visible on a world map. Lynxes are the only big cats that can be found on both the American and Eurasian continents.

Flamingo World Map

When you think of flamingos, the first places that come to mind are Florida or Africa. However, this map shows that flamingos are more widespread. Flamingos are native to numerous Mediterranean countries, the Caspian Sea and along the entire coast from Iran via Pakistan to India. The northernmost flamingo colony in the world is located in the Zwillbrocker Venn, a wetland in north-western Germany, which is home to a small group of 25 breeding flamingo couples every summer. By the way, the flamingo's famous one-legged stance appears to simply be a way to save energy. Because they can lock their joints, balancing on one leg requires much less muscle strength than using both legs.

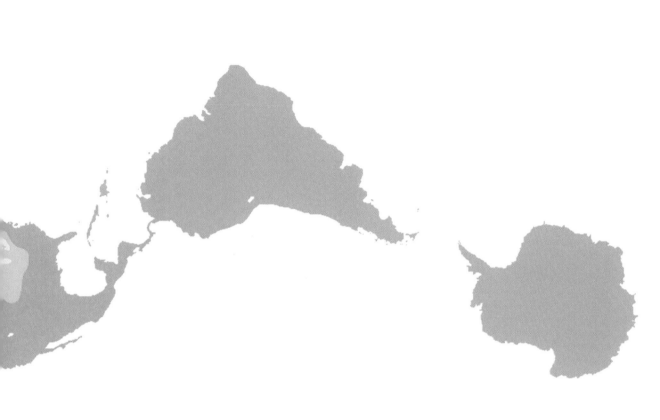

Pleistocene Mammoth Habitats

This map shows the habitat of an animal that is no longer with us — the woolly mammoth (Mammuthus primigenius). During the Ice Ages of the Late Pleistocene the light blue areas of the map were land bridges between the continents. While the oldest fossils of this species of elephant go back 800,000 years, they haven't been extinct for very long — the last mammoths lived around 4,000 years ago. At this point in time, the Great Pyramid of Giza had already been completed. Given how recent that is in geological terms, it's no surprise that in 2015 researchers were able to use a few specimens found in the Siberian permafrost to sequence the entire genome of the woolly mammoth.

Livestock Density

These maps show the population densities of livestock. Cattle are widespread everywhere because they make money — cash cows so to speak . . . Buffalo are mainly found in India and China. Of course, the high density of cattle and buffalo in India is also due to religious prohibitions. Goats can be found in poorer regions with rough terrain. Sheep were among the first animals that humans domesticated, so they are also widespread. Goats are most commonly found near mountains, because these areas are not only their natural habitat, but also difficult to cultivate otherwise. Pigs, chickens, and ducks are now mainly concentrated in China. They are a staple of traditional Chinese cuisine, and a large part of the population is now rich enough to consume meat on a regular basis. A mere 100 years ago, horses were used as a means of transport and farming around the world. With the arrival of the automobile this changed; today people keep horses mainly for recreational purposes.

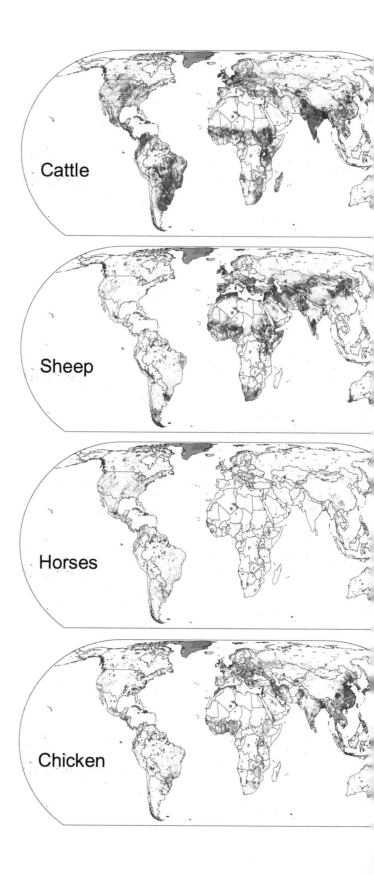

Cattle

Sheep

Horses

Chicken

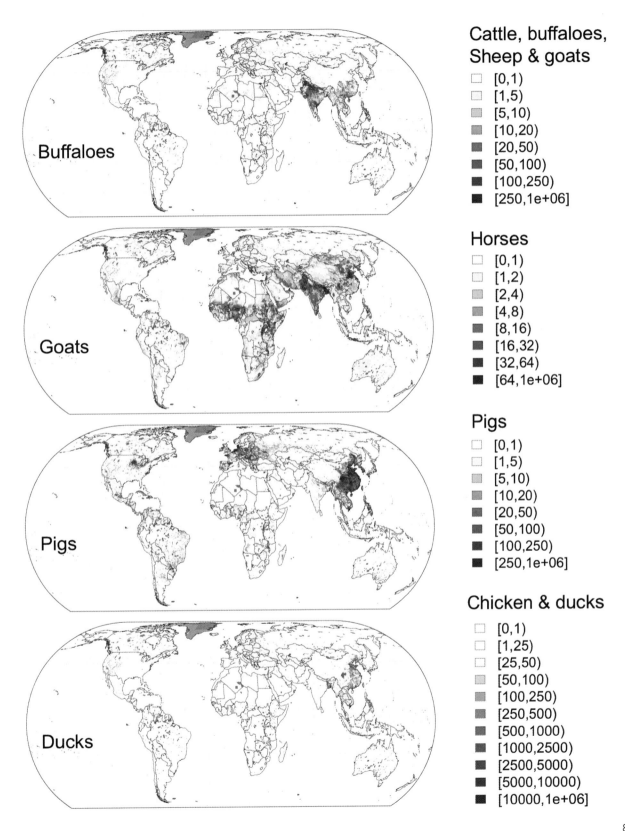

The World from a Dolphin's Point of View

We share our planet with many other living beings, and we are not the only intelligent inhabitants of this world. But we are the only species that can make maps. This inspired Dutch artist and designer Frans Blok to create a map from the point of view of dolphins. Most

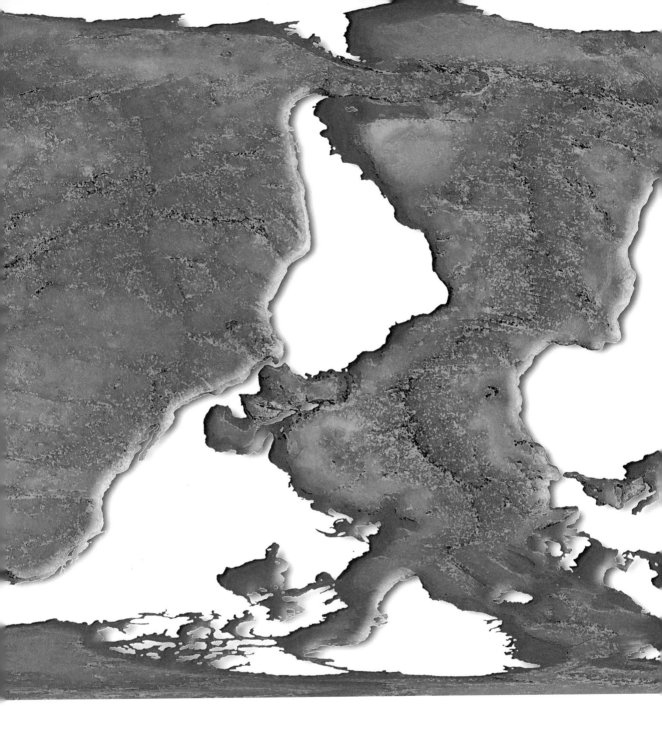

dolphins live in the Pacific, so this is the centre of the map. It shows the precise depths of the ocean. The map is upside down, since dolphins look out over the world from below. Unlike our maps, the dolphin map has many white spots left on it because the land masses remain terra incognita for these sea dwellers.

POLAR ANIMALS/WOLF

TWELVE ANIMALS/COCK

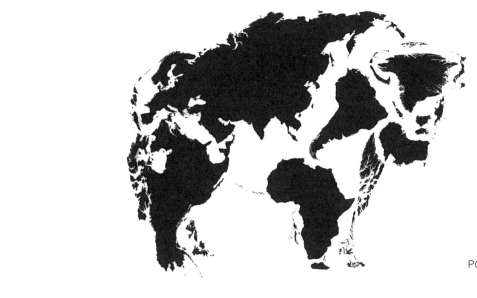

POLAR ANIMALS/BISON

POLAR ANIMALS/WHALE

World Animals

These maps show the familiar outlines of the continents and islands in a new way. The Japanese design agency Graflex Directions created this special collection of animals. The animal shapes are simply created through a new arrangement and rotation of the continents. The size of the individual continents and islands has not been changed. The design was occasioned by the Japanese custom of sending your friends, relatives, and business contacts, cards with traditional Japanese zodiac signs as New Year's greetings. The agency now works with various partners, including the WWF. Their creations are a way to wish the entire world peace and happiness with a map of the entire world. This gave the project its title, 'Piece Together for Peace'.

Number of Venomous Species per Country

This map shows how many venomous animal species can be found in each country around the world. The highest number is reported in Mexico (80), closely followed by Brazil with 79 species, while Australia has to contend itself with only 66 venomous animal species — giving

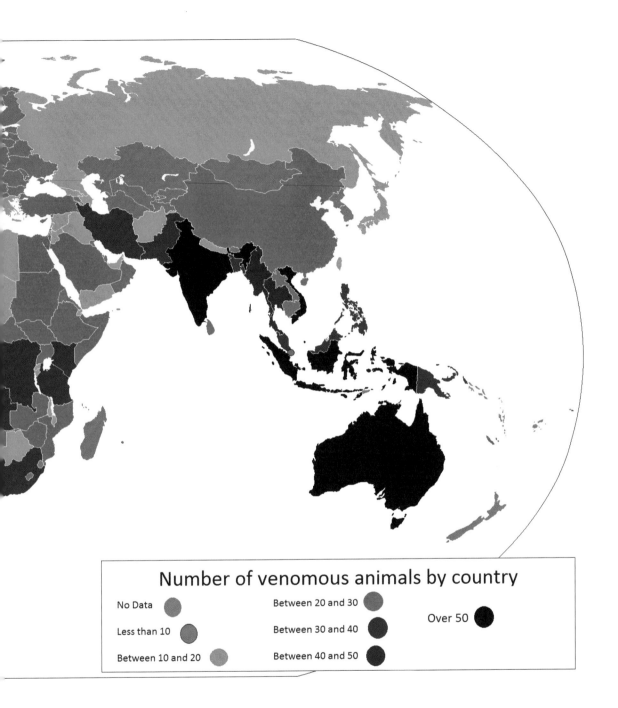

Number of venomous animals by country

No Data

Less than 10

Between 10 and 20

Between 20 and 30

Between 30 and 40

Between 40 and 50

Over 50

it the third place. This map forces biological distributions into the political boundaries of nation-states. Something like this may be a helpful and at times practical simplification, but it also leads to strange inaccuracies. France is red here because France also includes French Guiana on the north coast of South America.

The Extermination of the American Bison

Don't be fooled by this pink map — it's a display of humankind at its worst. When the pilgrim fathers arrived in what is now the United States, close to 30 million bison were estimated to occupy the interior plains. Bison was always hunted by Native Americans and was a never-ending food source — until the white man brought along his firearms and started decimating the bison population at an unimaginable scale. Hunting bison on horseback with rifles was easy, and soon enough everyone was doing it for fun, the once valuable corpses left to rot on the ground. In 1889, William Temple Hornaday of the New York Zoological Park predicted that the bison would be extinct within two decades. Hornaday went on to establish the American Bison Society in 1905, which, supported by Theodore Roosevelt, created bison sanctuaries to strengthen the population. The bison barely escaped extinction, which serves as a reminder that it is never too late to take positive action.

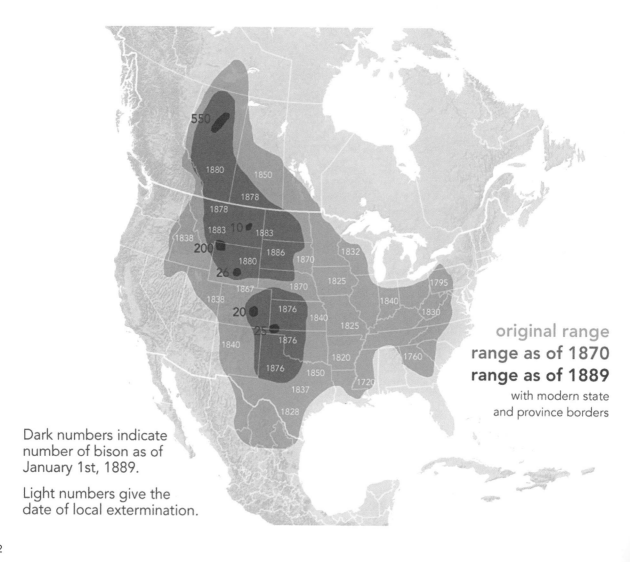

original range
range as of 1870
range as of 1889
with modern state
and province borders

Dark numbers indicate
number of bison as of
January 1st, 1889.

Light numbers give the
date of local extermination.

The Australian Dingo Fence

Animals are not always welcome guests. Australian sheep farmers, for example, are not particularly fond of dingoes. Sheep are not native to Australia, and the wild dogs gave the Europeans a lot of trouble, because for dingoes sheep were not only an exotic delicacy but also very easy prey. Since the 1890s, herders in Southeast Australia have been building larger and larger fences. In South Australia, the dingo fence became a matter of government policy with the Dog Fence Act of 1946. The fence is a continuous structure made of wire mesh, approximately 180 cm high with an additional 50 cm underground. In total, the fence is 3363 miles (5412 km) long — making the dingo fence the longest human-built structure in the world. Meanwhile, sheep farming in Australia has become a success story: 30 per cent of the global wool production comes from Down Under.

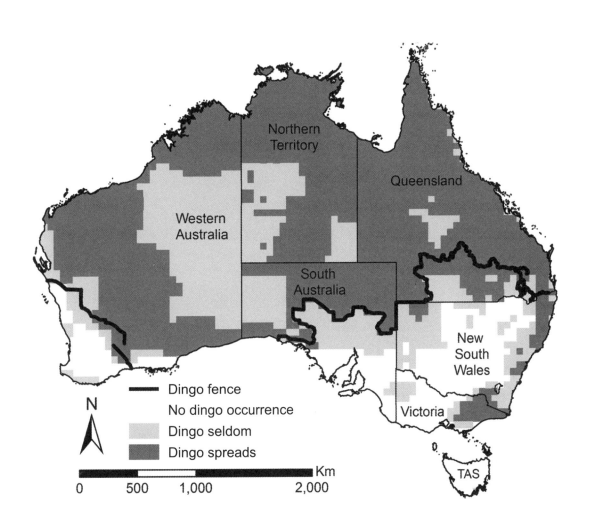

GEOGRAPHICAL TRIVIA

This chapter is a hodgepodge of wonderful geographical trivia. The term trivia refers to data, facts or circumstances without any particular scientific or practical use. I believe that not everything has to be useful. Besides, trivia is undeniably handy when it comes to small talk. If you have a repertoire of interesting facts, it becomes easier to chime into conversations. The next time a conversation turns to China, you might talk about the Heihe-Tengchong-Line, or, on the subject of South America, point out that it could just as easily be called East America.

In any case, the classification of information as trivia presupposes that useful knowledge can be objectively distinguished from useless knowledge. Because I don't believe that's true, in this chapter I present geographical oddities that I find subjectively interesting.

Pangaea with Modern Borders

This map takes us back to the distant past of our planet. 300 to 150 million years ago, the Earth's various continental plates were pushed together to form a single gigantic land mass — the supercontinent Pangaea. This continent is where the first dinosaurs evolved. This map shows the borders of present-day countries on Pangaea. It is, of course, enormously inaccurate, because back then a large part of the land we walk on today was still beneath the Earth's crust. But the map shows impressively how well the continents fit together — like pieces of a puzzle. It might also make you wonder how these countries would get along with their new neighbours. Some would even become landlocked, or, like Switzerland, gain sudden access to the sea.

1. Afghanistan
2. Albania
3. Armenia
4. Azerbaijan
5. Bangladesh
6. Belgium
7. Benin
8. Bhutan
9. Bosnia
10. Botswana
11. Brunei
12. Bulgaria
13. Burkina Faso
14. Costa Rica
15. Denmark
16. Germany
17. Ecuador
18. El Salvador
19. Ivory Coast
20. Eritrea
21. Estonia
22. Fiji
23. Finland
24. France
25. Gabon
26. Georgia
27. Ghana
28. Greece
29. Guatemala
30. Guinea
31. Guyana
32. Hawaii
33. Honduras
34. India
35. Indonesia
36. Iran
37. Ireland
38. Iceland
39. Israel
40. Italy
41. Jordan
42. Cameroon
43. Qatar
44. Kyrgyzstan
45. Congo
46. Croatia
47. Kuwait
48. Laos
49. Lesotho
50. Latvia
51. Liberia
52. Lithuania
53. Madagascar
54. Malawi
55. Malaysia
56. Marshall Islands
57. Macedonia
58. Micronesia
59. Myanmar
60. Nepal
61. Netherlands
62. North Korea
63. Norway
64. Austria
65. East Timor
66. Papua New Guinea
67. Philippines
68. Poland
69. Portugal
70. Romania
71. Sweden
72. Switzerland
73. Senegal
74. Serbia
75. Sierra Leone
76. Slovakia
77. Slovenia
78. Somalia
79. Spain
80. Sri Lanka
81. South Korea
82. South Sudan
83. Suriname
84. Swaziland
85. Syria
86. Tajikistan
87. Taiwan
88. Thailand
89. Togo
90. Czech Republic
91. Tunisia
92. Turkmenistan
93. Uganda
94. Ukraine
95. Hungary
96. Uzbekistan
97. United Kingdom
98. Vietnam
99. Belarus
100. Western Sahara
101. Central African Republic
102. Zimbabwe
103. Cyprus

Antipodes World Map

As children, many of us wondered: if I dug a hole straight down from where I am now, all the way through the centre of the Earth and out the other side, where would I end up? Now you have an answer to this question in the form of this map. Cartographers call this type of map an antipodal map, a term from Greek mythology that literally means a map of 'the people with opposite feet'. Our planet has only a few countries whose antipodes are also on land. Europe, Africa, and a lot of Asia meet the vastness of the Pacific, Australia's counterpart is the Atlantic and the antipodes of North America are mainly in the Indian Ocean. Some examples of country-to-country antipodes are Hawaii and Botswana, Argentina and China, New Zealand and Spain, or Chile and Russia.

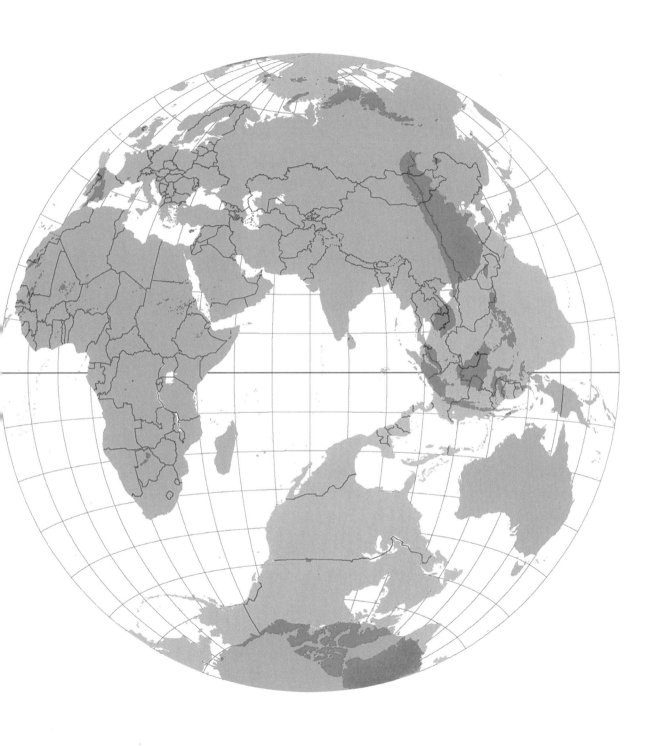

Date Line

A journey in a straight line (inasmuch as there are straight lines on our spherical planet) can have strange consequences, as this route from Hawaii to Antarctica shows. If you fly like this, you cross the date line no less than seven times. Today becomes tomorrow, then today again and tomorrow once more and today again and tomorrow again, today again and then finally tomorrow. This bizarre time travel is caused by the different countries to which the small islands and atolls of the Pacific belong. The unusual hook shape of the date line seen on the map is due to the Republic of Kiribati. This tiny nation with only about 110,000 inhabitants has a total land area of only 313 miles^2 (811 km^2), but it is enormous: 1275 miles (2051 km) from the northernmost to the southernmost island and a total of 2828 miles (4567 km) from east to west. As a result, the state has an area of 2 million miles^2 (5.3 million km^2), about half the size of the USA. And Kiribati also extends over both the equator and the 180th meridian — so it is the only country situated in the Northern, Southern, Western and Eastern Hemisphere at the same time. Originally, Kiribati straddled the date line and had two separate dates at once. At some point this became too challenging for the country's economy, which is why Kiribati unified its date for the whole nation on 1 January 1995 and now sits west of the date line, like Australia.

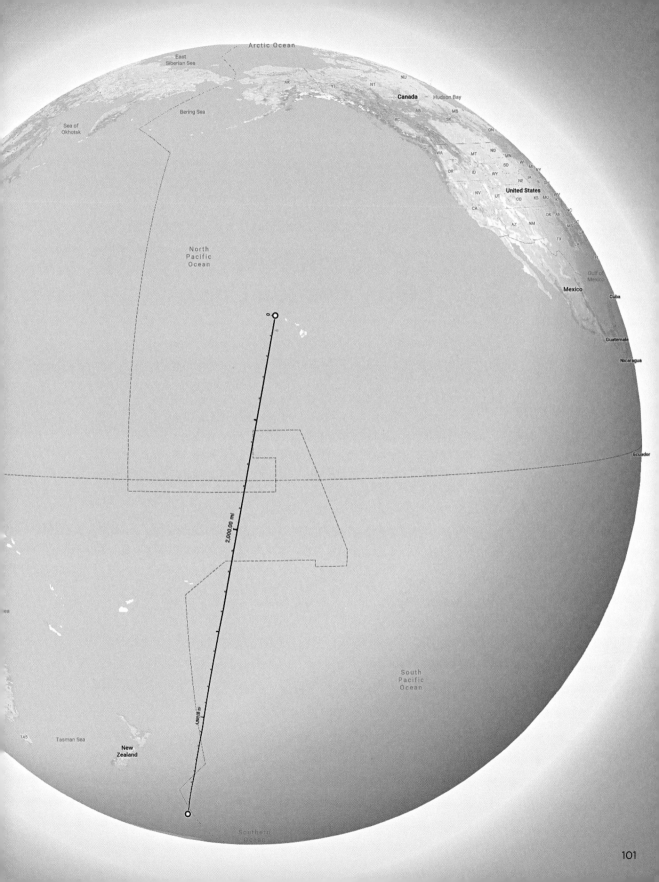

Canada Lives
Below the Red Line

Where should you emigrate to if you are looking for solitude? Canada would be a good choice — especially the northern part of the country. It is the second largest country in the world, but also one of the countries with the lowest population densities. A few large cities and metropolitan areas, such as Toronto in the South, are the exception. Half of all Canadians are concentrated in this small corner of the country, while the other 50 per cent are scattered across the vast expanses of the north and west. It gets loneliest in the far north, in the Nunavut region, which alone is sixteen times the size of England but only has as many inhabitants as the English town of Blyth in Northumberland (39,000 people). This results in a population density of 0.02 people per km^2.

80% of all Canadians live north of this line

80% of all Canadians live south of this line

Greenland is Farther East, West, North and South than Iceland

Sounds impossible, but it is true: Greenland is farther north, farther south, farther west and farther east than Iceland. That's because Greenland is the largest island on Earth — at least if we don't count Antarctica or Australia, since they are continents. That is why Greenland is much wider on the globe than Iceland. Incidentally, Greenland is also located in America, in the Arctic and in Europe — politically it is part of Denmark, geologically it is in the Arctic and geographically it belongs to North America.

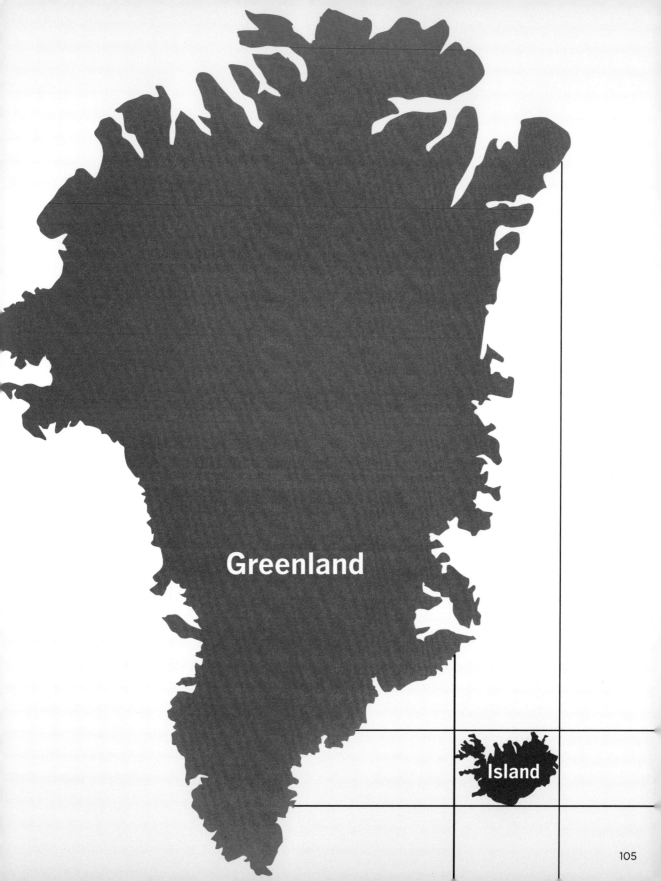

Greenland

Island

South America is also East America

From a US perspective, South America might just as well be called East America. The whole of South America is located east of the US state of Michigan and the city of Tampa in Florida. This fact is not really useful, but from now on it will be impossible not to mention it when talking about South America or Michigan.

Lo

CANADA

UNITED
STATES

Ottawa
★

Toronto
■

New York
■

★ Washington

MEXICO

Havana
★

BAHAMAS

CUBA

DOMINICAN
REPUBLIC

Kingston
★

HONDURAS

NICARAGUA

Caracas
★

PANAMA

Medellin
■

VENEZUELA

COLOMBIA

ECUADOR

BRAZIL

PERU

BOLIVIA

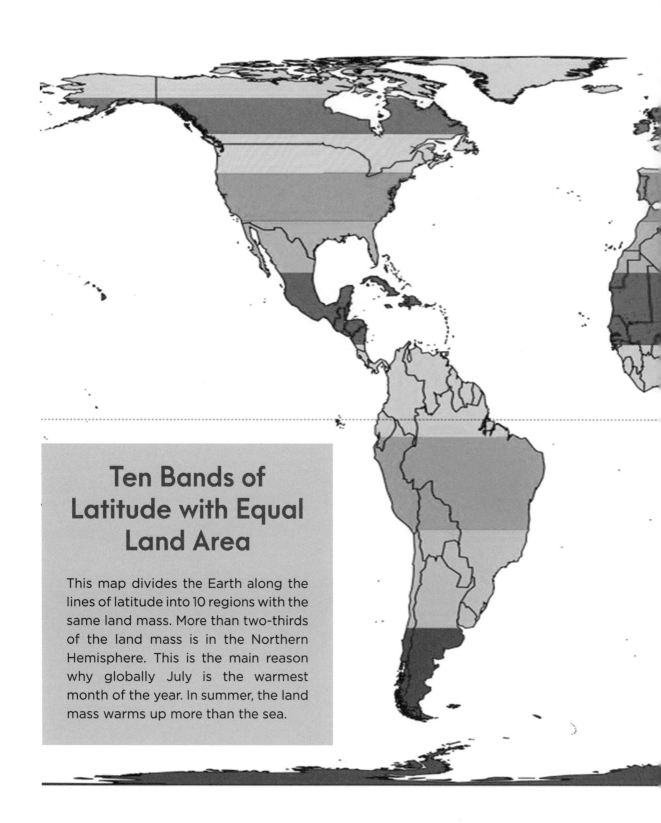

Ten Bands of Latitude with Equal Land Area

This map divides the Earth along the lines of latitude into 10 regions with the same land mass. More than two-thirds of the land mass is in the Northern Hemisphere. This is the main reason why globally July is the warmest month of the year. In summer, the land mass warms up more than the sea.

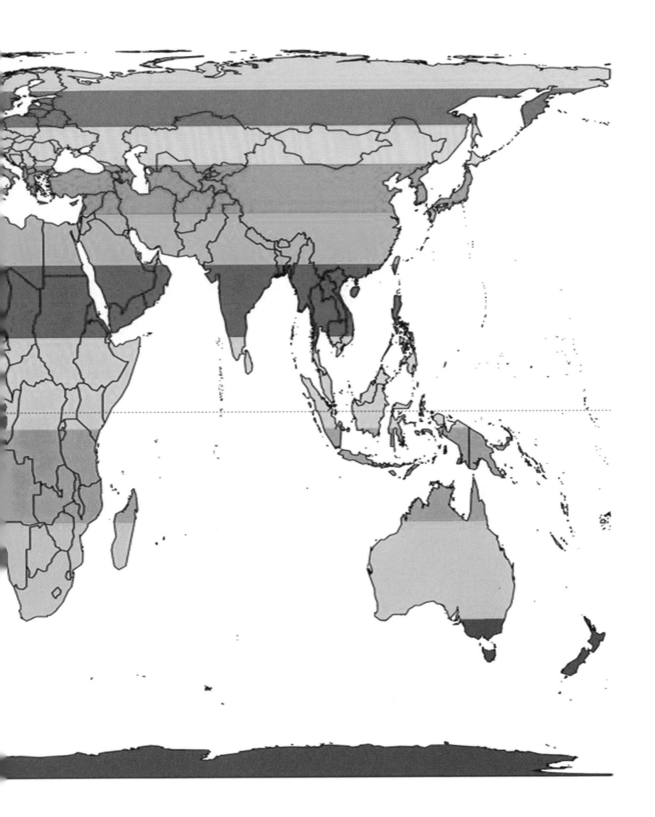

Australian Vegetation

This map of Australia was produced by the government in the 1920s to be published in textbooks or as a large-scale wall map to be displayed in classrooms. The map's creator, Thomas Griffith Taylor, simply wanted to educate students on the geography of their nation, and couldn't have possibly guessed that his work was to resurface a century later. The map became internet-famous for treating the island of Tasmania with bitter disrespect by labelling it as EW and WTF. Little could the cartographer know that the majority of readers wouldn't interpret WTF as Wet Temperate Forests...

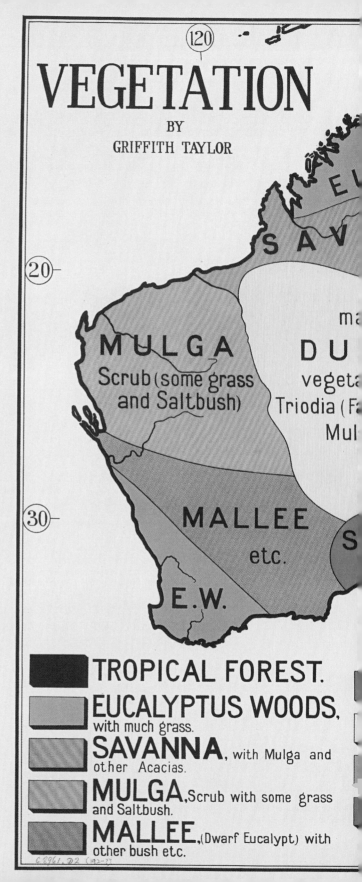

VEGETATION
BY
GRIFFITH TAYLOR

MULGA
Scrub (some grass and Saltbush)

DU
vegeta
Triodia (F
Mul

MALLEE
etc.

E.W.

TROPICAL FOREST.
EUCALYPTUS WOODS, with much grass.
SAVANNA, with Mulga and other Acacias.
MULGA, Scrub with some grass and Saltbush.
MALLEE, (Dwarf Eucalypt) with other bush etc.

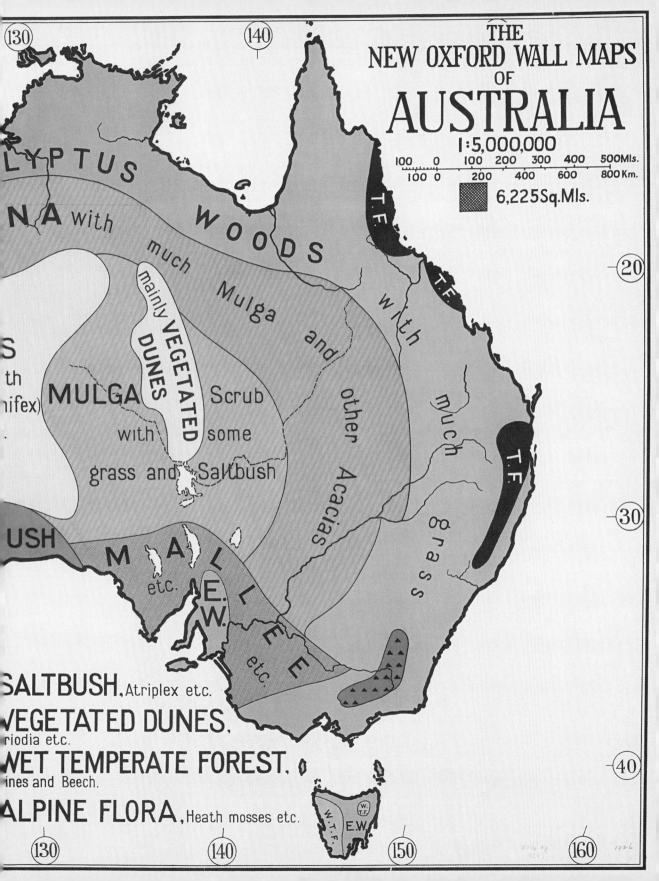

THE
NEW OXFORD WALL MAPS
OF
AUSTRALIA
1:5,000,000

100 0   100 200 300 400 500 Mls.
100 0   200   400   600   800 Km.

6,225 Sq. Mls.

LYPTUS

N A with

WOODS

much Mulga

mainly VEGETATED DUNES

MULGA

Scrub

with some

grass and Saltbush

MULGA and other Acacias

with much grass

T.F.

T.F.

T.F.

nifex)

USH

MALLEE etc.

etc.

E.W.

E.W.

SALTBUSH, Atriplex etc.

VEGETATED DUNES,
riodia etc.

WET TEMPERATE FOREST,
ines and Beech.

ALPINE FLORA, Heath mosses etc.

W.T.F.

W.T.F.

E.W.

T.F.

Heihe–Tengchong Line

The Heihe-Tengchong Line (also called the Aihui-Tengchong Line) is an imaginary line that divides the area of China into two roughly equal parts. This line runs diagonally across China, from Heihe in Heilongjiang Province to Tengchong in Yunnan Province. The western, yellow part of the map makes up 57 per cent of the area but houses only 6 per cent of the Chinese population. The red area east of the line takes up 43 per cent of the area and is inhabited by the other 94 per cent of the population. This distribution of the population is not a new development. The alluvial plains and coastal areas have always been densely populated. In the interior of China, on the other hand, there is an inhospitable and often downright hostile desert climate; this is where the Taklamakan and Gobi deserts are located. Also, the rapid population growth that China experienced in recent decades has occurred exclusively to the east of the Heihe-Tengchong Line.

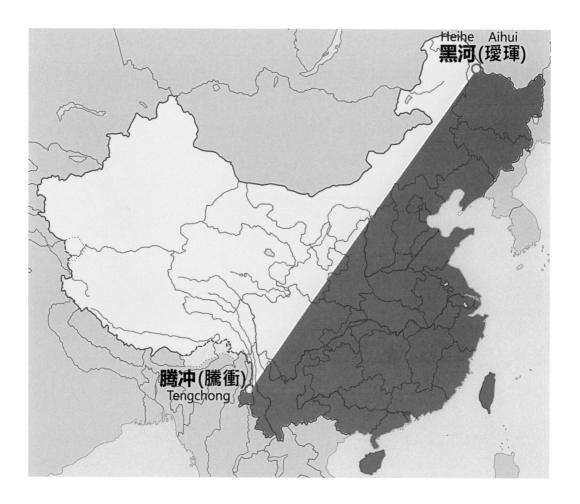

SIZE COMPARISONS

Size comparisons are a fun way to view our world from different angles. When we read something about large numbers, we can easily understand them without putting them into context. To understand data, it helps to compare it to something we know. This is data journalism 101: every number must be put in relation to a benchmark. These comparisons come in a variety of forms. If, for example, the population growth rate of Wales is reported, the national US average or at least the previous year's value for Wales can be used as benchmarks.

But clever journalists or cartographers sometimes come up with particularly creative and memorable comparisons. As a fan of maps, I especially enjoy visual aids. I often wish there were more of these, especially in daily newspapers or non-fiction books. But since these visual aids are usually created by designers or cartographers and not the journalists and authors themselves, their scarcity is not surprising. As is so often the case, the failure to include truly helpful size comparisons is caused by time constraints and limited resources. This is why I was all the more pleased by the examples on the following pages.

Mercator Projection

Perhaps the most important size comparison map was created by one of the most influential cartographers in the online mapping community: Neil Kaye, who works for the UK Weather Service. The original Mercator Projection remains a hugely popular map. But it magnifies regions that are far from the equator significantly more than regions close to the equator. Kaye's map compares the Mercator Projection (light blue) with the actual size of countries (dark blue). It becomes clear that the Mercator results in an inflated Russia, a monstrous Greenland and a somewhat enlarged Europe. Most cartographers cut Antarctica off from Mercator projections. This map, however, impressively shows the distortion of the South Pole region.

The True Size of Africa

For many people in the West, Africa is probably the largest gap in their geographical knowledge. This map tries to fill the gap a little by showing how big the African continent really is. On Mercator maps, the landmasses north of Africa are distorted and look larger than they are. That in turn makes Africa look much smaller. By filling in the outline of Africa with landmasses that are perhaps more familiar to us we get a better understanding of the true size of the continent.

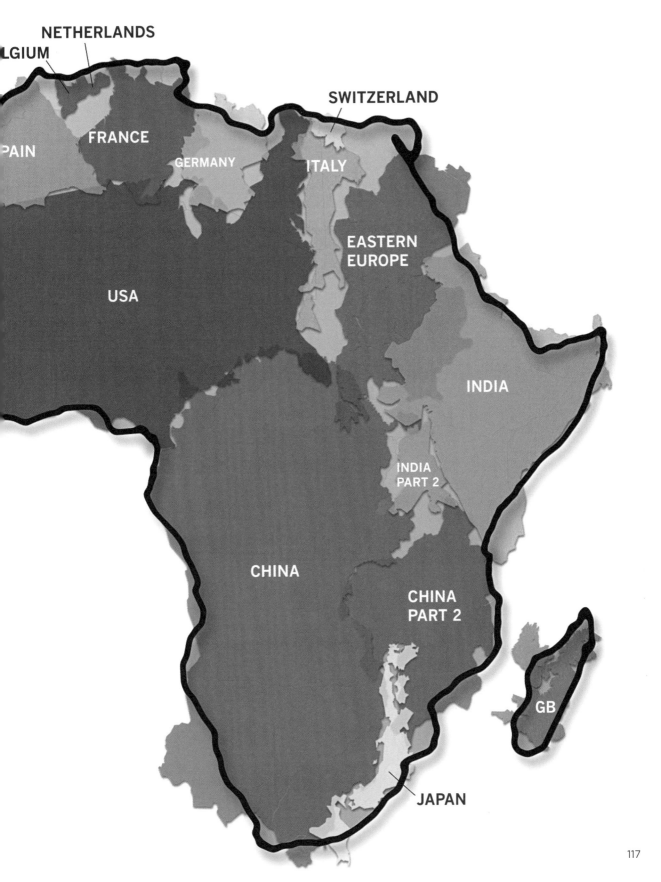

Rivers and Lakes of the World

One of the most spectacular size comparison maps was created by James Reynolds in London in 1851. Rivers and lakes from all over the world are shown here side by side — even important cities along the rivers are included. It is immediately apparent that even the mightiest streams of Europe, such as the Danube, are dwarfed by numerous rivers on other continents. The map is now almost 200 years old, so it is not surprising that some physical realities have changed dramatically. For example, the Aral Sea (number 17) has now almost completely dried up and would probably no longer be found on a new edition of the map.

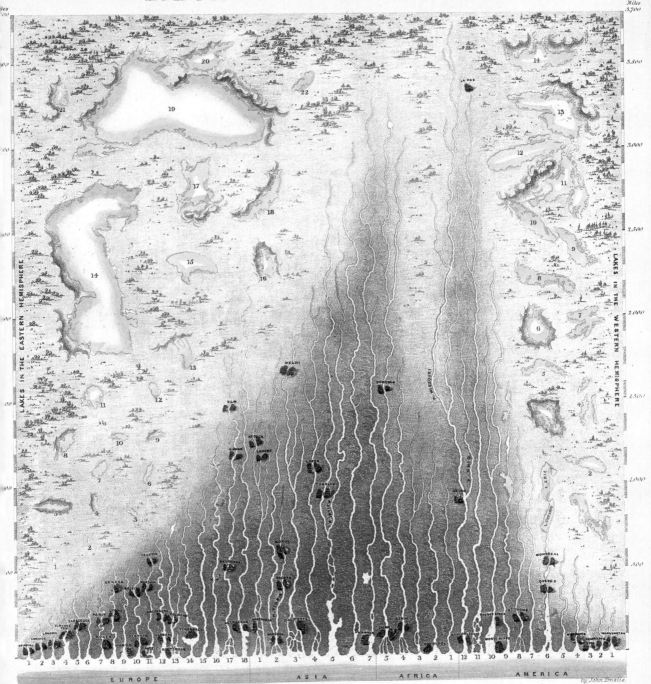

Drawn & Engraved

by John Emslie.

EUROPE

ASIA

AFRICA

AMERICA

RIVERS.

EUROPE.	ASIA.
Locality	7 Yang tse, *China*
1 Tay, *Scotland*	6 Po, *North Italy*	12 Elbe, *Germany*	1 Indus, *Cabul &c.*
2 Shannon, *Ireland*	7 Seine, *France*	13 Rhine, *Germany*	2 Euphrates, *A Turkey*
3 Thames, *England*	8 Rhone, *France*	14 Don, *Russia*	3 Ganges, *Hindostan*
4 Severn, *England*	9 Tagus, *Spain and Por.*	15 Dwina, *Russia*	4 Burrampoota, *Tibet*
5 Ebro, *Spain*	10 Oder, *Prussia*	16 Dnieper, *Russia*	5 Obi, *Siberia*
	11 Vistula, *Poland*	17 Danube, *Germany*	6 Hoang ho, *China*
		18 Volga, *Russia*	

AFRICA.

1 Orange, *Namaqualand*
2 Gambia, *Senegambia*
3 Senegal, *Senegambia*
4 Niger, *Nigritia*
5 Nile, *Egypt & Abyssinia*

AMERICA.	
1 Potomac, *U. States*	7 Bravo del Norte, *Mexico*
2 Delaware, *U. States*	8 Oronoco, *Columbia*
3 Hudson, *U. States*	9 M.Kenzie, *Indian Ty.*
4 Susquehana, *U. States*	10 La Plata, *La Plata*
5 Colorado, *La Plato*	11 Amazon, *Brazil*
6 St. Lawrence, *Canada*	12 Mississippi, *U. States*

LAKES.

EASTERN HEMISPHERE.			
Surface Sq. Miles	5 Zaizan, *Mongolia* 1600	11 Ladoga, *Russia* 5200	17 Aral Sea, *Tartary* 11650
1 Lough Neagh, *Ireland* 80	6 Wetter, *Sweden* 945	12 Dembia, *Abyssinia* 1300	18 Baikal Sea, *Siberia* 8000
2 Loch Lomond, *Scotl.* 27	7 Enare, *Lapland* 870	13 Orega, *Russia* 3300	19 Black Sea, *Turkey* 113000
3 Constance, *Switzerland* 456	8 Toming, *China* 1200	14 Caspian Sea, *Russia* 119000	20 Azov Sea, *Russia* 8800
4 Dead Sea, *Syria* 370	9 Geneva, *Switzerland* 400	15 Tchad, *Africa* 11600	21 Balkash, *Mongolia* 5200
	10 Orebo, *Sweden* 900	16 Ouroomia, *Persia* 900	22 Wener, *Sweden* 2100

WESTERN HEMISPHERE.		
Surface Sq. Miles	5 Athabasca, *N. America* 3200	10 Erie, *North America* 4800
1 Winnepagos, *N. Amer.* 2000	6 Maracaibo, *S. America* 6000	11 Huron, *North America* 15800
2 Nicaragua, *N. America* 2905	7 Great Bear, *N. America* 4000	12 Michigan, *N. America* 12600
3 Titicaca, *S.th America* 5400	8 Winnepeg, *N. America* 7200	13 Superior, *N. America* 22400
4 Otehenantekane, *N. Am.* 2500	9 Ontario, *N. America* 4450	14 Great Slave, *N. Amer.* 12000

London Published by James Reynolds 174 Strand.

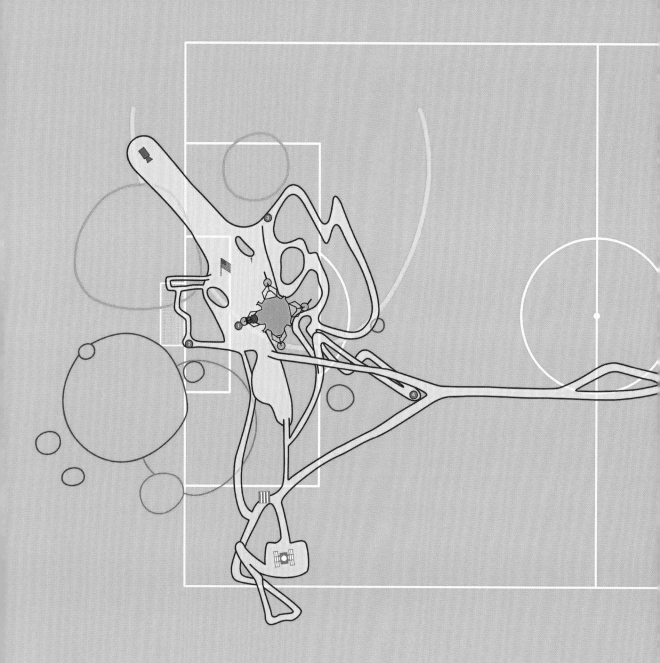

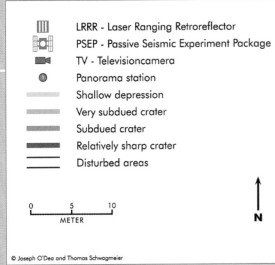

Football with Neil and Buzz

In 1969, the Apollo 11 mission took Neil Armstrong and Buzz Aldrin to the moon. One giant leap for mankind, but only one small step for a man? Neil and Buzz's moonwalk seems pretty manageable once projected onto a football (soccer) field. The two astronauts only spent 2 hours and 31 minutes on the moon's surface, an hour longer than a normal football game. But Neil and Buzz only made a brief foray into the opponent's half of the pitch.

IIII	LRRR - Laser Ranging Retroreflector
	PSEP - Passive Seismic Experiment Package
	TV - Televisioncamera
①	Panorama station
	Shallow depression
	Very subdued crater
	Subdued crater
	Relatively sharp crater
	Disturbed areas

0   5   10
METER

N

© Joseph O'Dea and Thomas Schwagmeier

Overlay of the Roman Empire and the US

This map makes an unusual size comparison between present-day US and the Roman Empire at the time of its greatest expansion. The red area is the Roman Empire in 117 AD, the year of the death of Emperor Trajan. The outline was only shifted in terms of longitude, i.e. horizontally, but not vertically — which once again shows that Rome and New York are less than one degree of latitude apart.

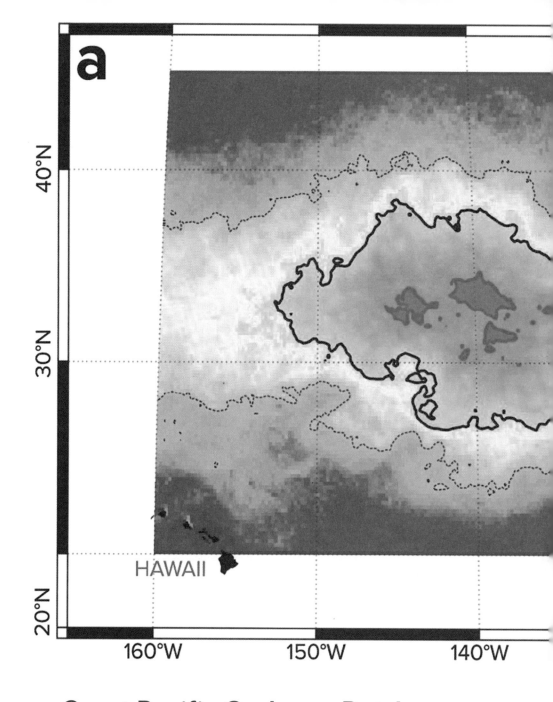

Great Pacific Garbage Patch

Of course, size comparisons can also be political. An example for this is the accumulation of plastic waste in the Pacific, the so-called 'Great Pacific Garbage Patch'. This map from the prestigious science magazine Nature shows the Great Pacific Garbage Patch between Hawaii and California. It allows us to grasp the area's massive dimensions. The problem here is that you have to read the map legend very carefully. It tells us how much plastic is floating on the

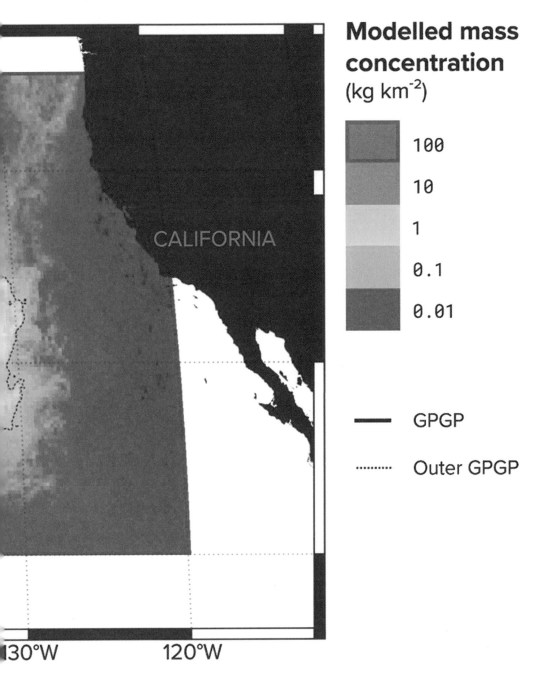

Modelled mass concentration (kg km^{-2})

	100
	10
	1
	0.1
	0.01

—— GPGP

.... Outer GPGP

CALIFORNIA

130°W   120°W

surface of the water per square kilometre, i.e. how high the plastic density is. One hundred kilograms per square kilometre is an alarmingly high figure. But since the Great Pacific Garbage Patch is not a cohesive, solid garbage island, critics see the map as exaggerated scaremongering. Inadvertently, this popular size comparison is used as evidence, not only by environmentalists wanting to draw attention to plastic pollution, but also by critics looking for material to portray these environmentalists as alarmists.

CULINARY DELIGHTS

All over the world, people are rightfully proud of their local delicacies. So much food, so little time. While we can't get a taste of everything, we can at least enjoy a few low-calorie maps.

The maps in this chapter all revolve around food, but they reveal a lot more. How do climate zones, trade routes, and wealth affect our eating habits? How far did our favourite drinks have to travel to get to us? On the next few pages, you will find a few of the most interesting culinary maps. On the internet, you can find many more of these, especially for beer and coffee. Here, I present only a small selection, hoping that this will whet your appetite for more.

Coffee and Wine Belts

Coffee and wine are definitely among the most popular drinks in the world. Both drinks are plant-based, and the cultivation of these plants requires great expertise. The map shows the wide belt around the equator — from around 23.5 degrees north to 23.5 degrees south —

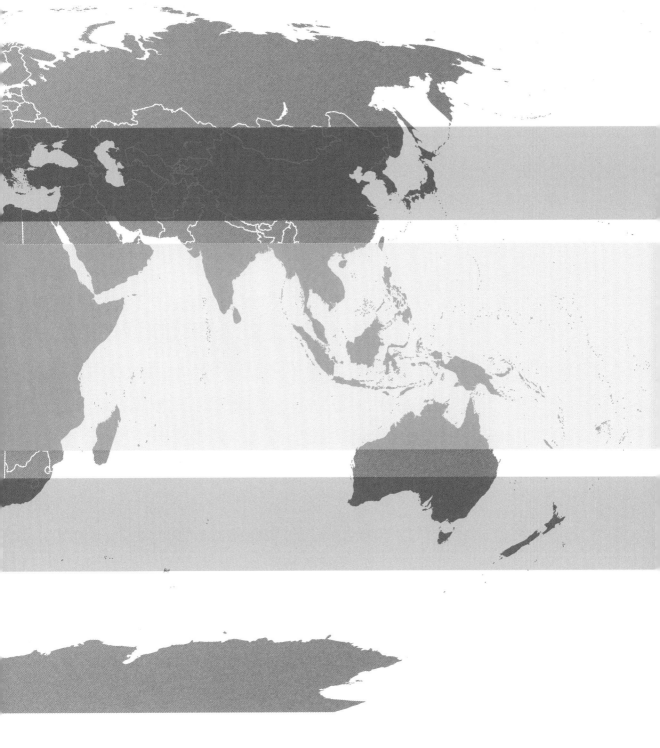

where coffee plants feel at home. Meanwhile, grape vines require a less tropical climate. They prefer the latitudes 30 to 50, either to the north or south. Thus, a map of our drinking habits becomes a map of global trade routes.

Tongue-in-Cheek Alcohol Preferences of Europe

This map is more of a joke at the Hungarians' expense than serious data analysis. Allegedly the Hungarians drink everything, so long as it is alcoholic. Of course, this is not really the case. Nevertheless, the map shows the drinking preferences of Europeans accurately in some places: the British prefer beer, Southern Europeans drink wine and Eastern Europe enjoys vodka. But the presentation is so simplified that it leads to a few amusingly wrong conclusions. If you believe the map, Bavarians enjoy wine just as much as beer and the Irish are — at least partially — completely abstinent. This is because a Venn diagram is not suitable for representing geographical distributions. Venn diagrams visualise mathematical intersections. The two sets of numbers are usually represented as circles or ellipses, but political and cultural boundaries are rarely so simple and elegant. A more accurate representation of the overlaps in the European drinking preferences would look like it had been drawn by a drunk.

Actual Alcohol Preferences of Europe

Here we have a more serious map of European drinking preferences. It is based on the yearly average consumption of alcohol per person. The upper map shows the preferences in 1990, the lower map the results for 2015. Immediately we see the massive changes that took place: the Russians, who once had a strong preference for spirits (obviously we are talking about vodka here), are now drinking more beer. The Spaniards have not remained loyal to wine. Meanwhile, in Germany and Great Britain the preference for beer is waning, and Swedes and Danes have discovered wine. Not all these developments can be traced back to cultural trends, or the effects of globalisation or even the gradual unification of Europe. In Iceland, for example, beer was simply illegal until 1989 while spirits remained legal. Eventually the government changed its stance, now believing that letting people drink beer is a better way to prevent alcoholism. The numbers on this map back them up, because once beer was legalised, Icelanders consumed less hard liquor. The high price of alcohol in Scandinavia may also explain why Estonia has remained such a bulwark of alcohol consumption — many tourists from Scandinavia stock up on the comparatively cheap alcoholic drinks there.

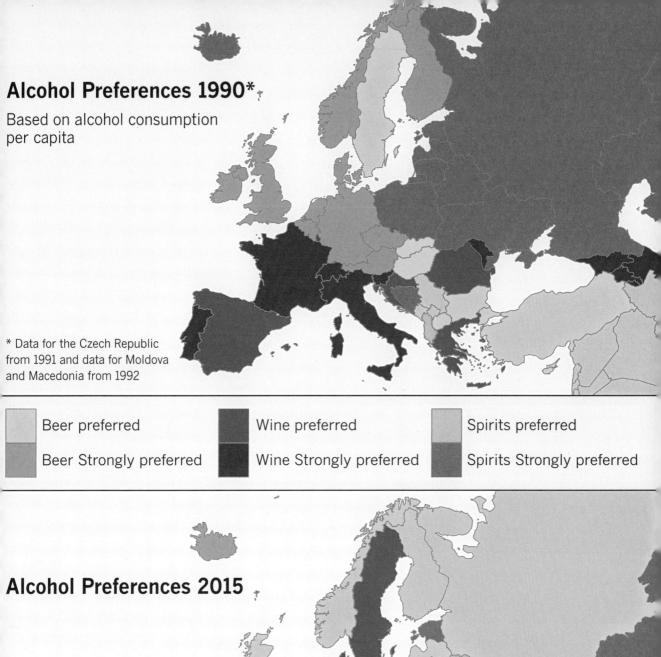

Alcohol Preferences 1990*

Based on alcohol consumption
per capita

* Data for the Czech Republic
from 1991 and data for Moldova
and Macedonia from 1992

Beer preferred	Wine preferred	Spirits preferred
Beer Strongly preferred	Wine Strongly preferred	Spirits Strongly preferred

Alcohol Preferences 2015

Obesity in Europe

This map shows the percentage of the population in each European country that is considered obese. Where do the fattest Europeans live? According to this map, Turkey outweighs Great Britain, Hungary, and Lithuania at 32 per cent. This does not mean, however, that Turkey is home to the largest individuals, just the largest number of obese people. Bosnia has the lowest rate of obesity, but there are still more than enough overweight people there. Incidentally, the medical term 'obese' refers to people whose body mass index (BMI) is above 30. On the other hand, anyone above a BMI of 25 is overweight. The numbers for England are less than flattering for example. About 28 per cent of the population are classified as obese while a further 36 per cent are classified as overweight. This means that about two-thirds of the English population are either overweight or obese.

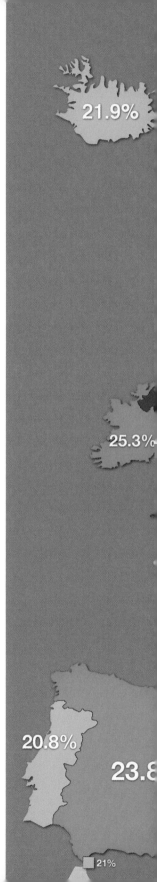

21.9%

25.3%

20.8%

23.8

21%

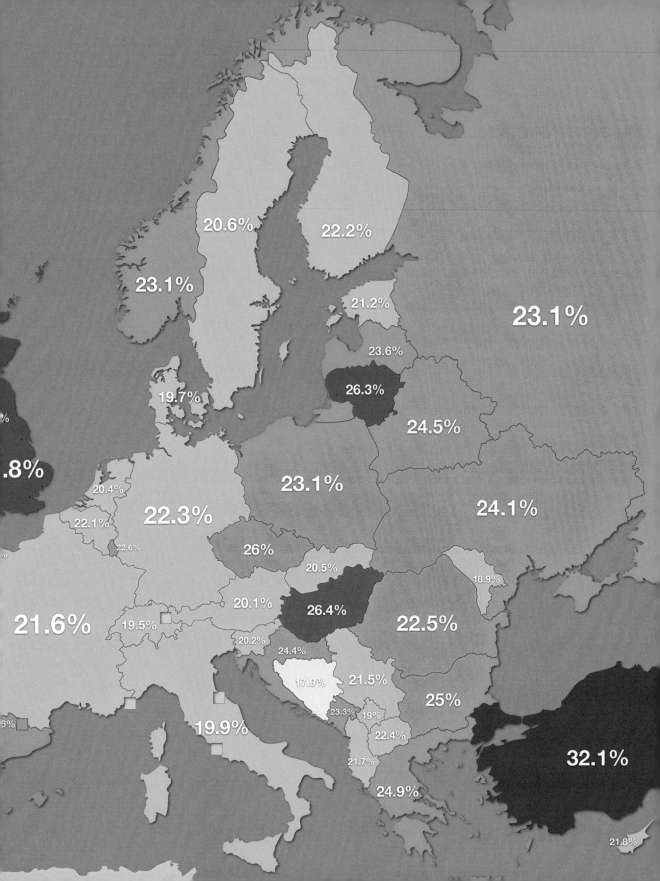

EUROPE

As a German who lived in the US for a while and is now based in Australia, I am very used to being labelled as a European rather than a German. It does indeed require a few years living outside of Europe to see that there is more to glue Europe together than divide it into distinctly different cultural spheres.

That said, Europe is wonderfully diverse and thus there are countless maps that display this diversity, especially when it comes to languages. Other maps look at business and the arts. We will see what connects Europe and where we are most likely to notice differences between us and our neighbours.

Icelandic Krona

Currencies Before the Euro

What were the currencies of the EU member states called before they adopted the Euro? This knowledge is of importance for both economic historians and crossword enthusiasts. In Montenegro, the currency was technically the Deutschmark. When migrant workers returned from Germany, they brought along their savings in German currency and the mark became an unofficial means of payment in former Yugoslavia. After the war in 1997 the convertible Marka was then fixed one-to-one against the Deutschmark. Although Montenegro is not actually a member of the EU, it did adopt the Euro when it was introduced.

Irish Pound

■ Member countries of the euro

■ EU member states, not euro area member states

■ Non EU member countries

Portuguese Escudo

Spani

- ■ **Mother tongue English**
- ■ **Duration about 24 weeks**
- ■ **Study time about 30 weeks**
- ■ **Study time approx. 44 weeks**
- ■ **No specific learning duration, but longer than 44 weeks**

Difficulty of European Languages

How hard is it for native English speakers to learn various foreign languages? The US government's Foreign Service Institute has been teaching foreign languages to diplomats for many decades and has collected enough data to determine the answer to this question. German is harder to learn than Italian, Spanish and French, but much easier than Eastern European languages. English-speaking diplomats have to invest even more time if they want to learn Hungarian or Finnish. If the foreign language also has an unfamiliar alphabet, then it takes even longer to learn. Arabic or Chinese (not shown on the map) are particularly difficult.

Blue Banana

The port of Rotterdam is located right at the centre of the 'Blue Banana'. This quirky term refers to the banana-shaped region of high population density and economic power that stretches from Liverpool to Genoa all the way through Central Europe. The French economic geographer Roger Brunet coined the term in 1989, when trying to describe the European backbone of industry and services that stretches from Northern England along the Rhine to Northern Italy. The blue banana developed out of historical trade routes and the geographical distribution of early capital flows. The concept of the blue banana has often been criticised for neglecting the greater Paris area, which has a considerable impact on the European economy. Why is the banana blue? That's for us to decide. It is either an allusion to the European flag (blue with yellow stars) or a reference to blue collar industrial workers.

Manchester

Liverpool

Birmingham

London

Bristol

Amsterdam

Rotterdam

Essen

Bonn

Brüssel

Luxemburg

Frankfurt

Stuttgart

Strasbourg

Zürich

Geneva

Turin

Milan

Genoa

What Would Other Brexits Be Called?

The portmanteau 'Brexit' is now firmly anchored in our vocabulary. Many Europeans might find the idea that other countries could follow the British example worrisome. Nevertheless, we can turn it into a fun intellectual game: what would other countries call their exit from the EU? This map offers some suggestions.

Sp

Abortugal

Portugo

Departugal

Number of Letters in European Alphabets

What at first glance might look like a ranking in which Slovakia does particularly poorly within Europe, is really the number of letters in the European alphabets.

But shouldn't Germany have more letters than Great Britain (both 26)? After all, German has three umlauts and an extra S (ß). In fact, depending on how you define letters, German has up to 30 different ones. It turns out that finding a consensus on this isn't easy — at least for linguists. Umlauts are letters with little extras, so-called diacritical marks, and the Slovak alphabet (with its record 46 letters) has 17 of them! The Slovaks use the same Latin alphabet as the Germans or Italians, but they have decorated it with a whole number of dots, ticks and arcs. And what is going on with the Italian alphabet? Why are there only 21 letters here? It is very close to the Roman original and today's Italians usually only use the letters J, K, W, X and Y to spell foreign words, so these letters are considered extensions. The Irish are even more economical: Irish Gaelic manages without J, Q, V, W, X, Y or Z.

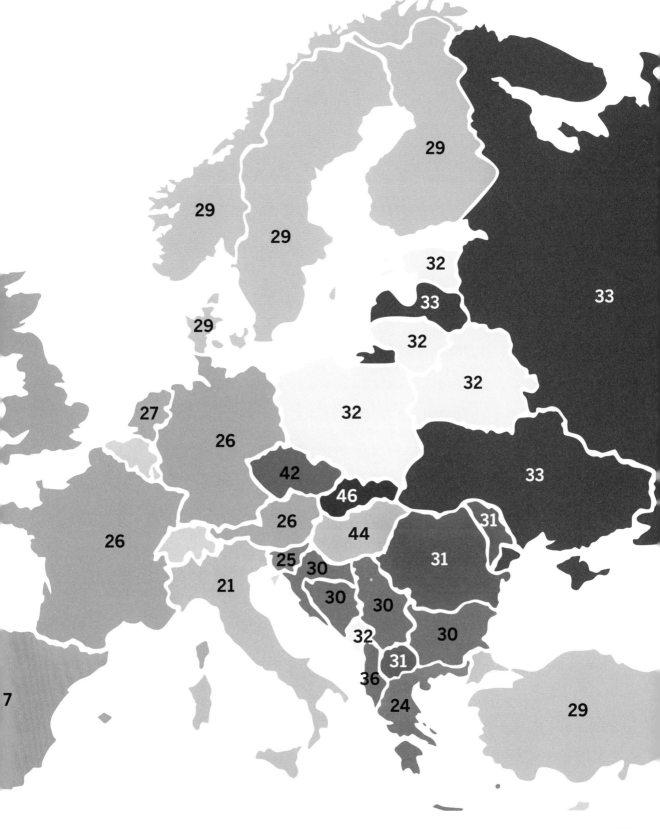

The Surname 'Smith' in European Languages

Your surname is a gateway into family history and etymology. Many family names go back to a profession that some ancestor once practiced. Mr or Mrs Smith can therefore assume they have a blacksmith in their family tree.

In the Middle Ages, blacksmith was a common and respected profession. Many different varieties of smiths existed — blacksmiths, fine smiths, bladesmiths, goldsmiths, armourers, coppersmiths and so on.

The map shows the surname 'Smith' in European languages. The colours of the regions indicate the common origins of the variants. The names in red areas all come from the Germanic 'smiþa'. In blue areas, the names have their origin (as befits Romance languages) in the Latin word 'ferrum' (iron). The proto-Slavic word 'kovati' describes the process of forging and led to surnames such as Kowalski (Poland) or Kovalyov (Russia). Of course, each spelling of the surname on this map is only one of many possibilities. In German, for example, there's not only Schmidt but also Schmitt, Schmied or Schmitz, as well as many other surnames such as Eisenhauer or Faber, which are based on the same group of professions.

Mac Ga

F

Ferre

Smed

Seppänen

Sepp

Ковалёв
Kovalyov

Kalējs

Kalvaitis

Smed

Smed

Кавалёў
Kavalyow

an

Smith

Smit

Kowalski

Schmidt

Schmit

Kovář

Kováč

Коваленко
Kovalenko

Goff

Kovač

Kovács

Lefebvre

Kovač

Feraru

Arotza

Ferrari

Ковач
Kovač

Ferrer

Ковачев
Kovachev

Ковачев
Kovačev

Nallbani

rero

Σιδεράς
Sideras

Demirci

149

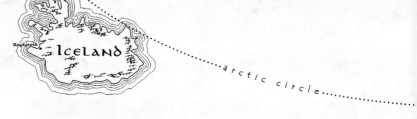

Fantasy Map of Europe

This map shows Europe in the style of J. R. R. Tolkien's Lord of the Rings. Peter Bird, professor of geophysics at the University of California, compared the topology of Middle-Earth and Europe and arrived at the conclusion that Mordor is in Romania. And not just anywhere in Romania, but in legendary Transylvania. The Shire, the land of the hobbits, is, according to Tolkien himself, a romanticised version of rural England. The movies were of course all shot in New Zealand.

French Kissing

How many kisses are there in a French greeting? This map by Bill Rankin, one of the most original living cartographers, is a classic. Instead of simply showing the most frequent answer per regional department and thus making it a boring and somewhat imprecise map, Rankin reached deep into the bag of tricks. The map shows all the answers for each regional department as individual coloured points. This allows it to reveal some of the more intricate details of French greeting habits. On the island of Corsica, according to 18 per cent of the participants, people commonly greet each other by kissing five times.

Nantes

Bo

One

Two

Three

Four

Five

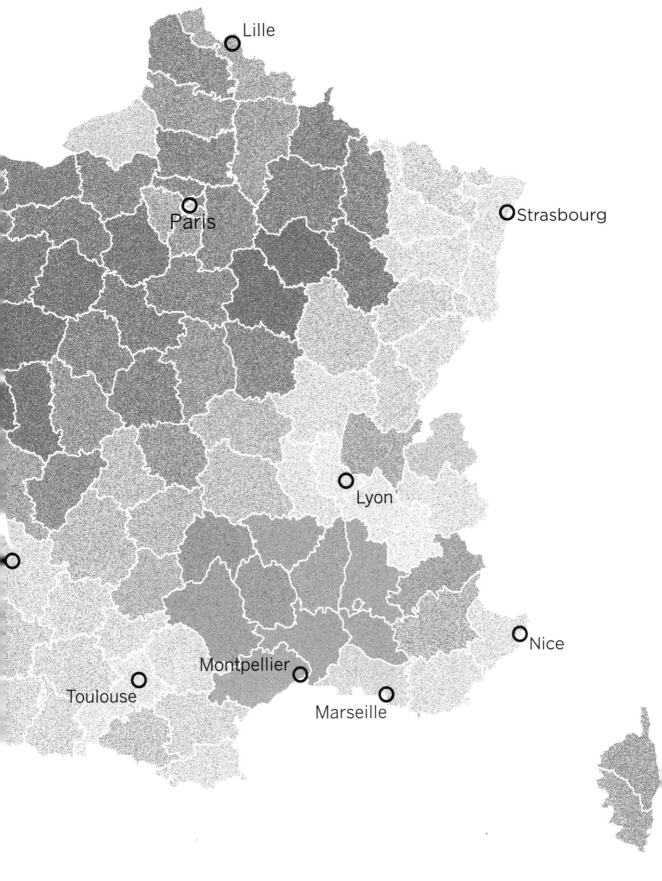

Who Brings the Christmas Presents?

This map by Jakub Marian is about Christmas presents. More precisely, the beliefs about who delivers these presents. In some countries there are a few competing traditions. Germans, for example, either believe in Santa Claus or the Christkind (Christ Child), hence the colour gradient on the map. And the whole of Europe is colourfully diverse in this regard. In the Soviet Union, Grandfather Frost brought the presents. This custom has been preserved in many countries of the former Eastern Bloc, but at the same time Christian traditions are making a comeback with St. Nicholas or Baby Jesus as gift-bringers. In Spain, gifts are traditionally delivered by the Three Wise Men on January 5th or 6th. But thanks to the ubiquity of the 'American' Santa Claus, some Spaniards now receive their gifts from 'Papá Noel' on Christmas Eve.

Joulupukki
literally a Christmas goat, nowadays conflated with a character similar to Santa Claus

Jultomten
Christmas gnome (mostly represented by an old bearded man)

Julgubben
Christmas Old Man (Swedish speakers)

Julenissen
Christmas gnome (mostly represented by an old bearded man)

Jõuluvana
Christmas Old

Ziemassvētku vecītis
Christmas Old Man

Nollaig
Old Man

Kalēdų Senelis
Grandfather Christmas

Дед Мороз
[Ded Moroz]
Grandfather Frost (Old Man Frost)
brings presents on New Year's Eve or New Year's Day rather than Christmas

Julemanden
Christmas Man

Дзед Мароз
Grandfather Frost (Old Man Frost)

Father Christmas

Kerstman
Christmas Man (a recent invention; traditionally no Christmas gift-bringer)

Weihnachts-mann
Christmas Man

Święty Mikołaj
Saint Nicholas
or **Gwiazdor**
Star-man
or **Aniołek**, *Angel*
or **Dzieciątko**
Jesus Child

Святий Миколай
Saint Nicholas on 19 December

(many Ukrainian families celebrate both)

Дід Мороз
Grandfather Frost (Old Man Frost) on New Year's Eve

born
nney

Ježíšek
Baby Jesus

Ježiško
Baby Jesus

Christkind
Christ-child

Père Noël
Father Christmas

Jézuska (+**Angyal**)
Baby Jesus (+**Angel**)
(or Télapó, a recent import)

Angyal (*Angel*)
(Hungarian speakers)

Moș Crăciun
Old Man Christmas

Božiček
Christmas Man
or **Dedek Mraz**
Grandfather Frost

Djed Božićnjak
Grandfather Christmas

Deda Mraz
Деда Мраз
Grandfather Frost

Дядо Коледа
Grandfather Christmas
or **Дядо Мраз** *Grandfather Frost*

Tió de Nadal
Christmas log (tree trunk that defecates presents)

Babbo Natale
Daddy Christmas (based on foreign traditions)
or **Gesù bambino**, *Jesus child (traditional in some regions but now uncommon)*
or **la Befana** (*a witch*) on 5 January
or **Santa Lucia**, *Saint Lucy* on 13 December

Дедо Мраз *Grandfather Frost*

Babagjyshi i Vitit te Ri
Grandfather of the new year

Noel Baba
Christmas Father (for non-Christians a general symbol of the New Year)

Pare Nadal
Father Christmas

OS

ed

Ἅγιος Βασίλης
Saint Basil (traditionally on the 1st of January)

Father Christmas
(Borrowed from the British)

What is Germany Called in Other European Languages?

The map shows the different names for Germany in Europe and North Africa. You might say that most neighbouring countries simply named Germany after the Germanic tribe that launched the most frequent attacks on them. In fact, the diversity is due to the long and inconsistent history of the Germanic peoples. After all, Germany was still a disparate collection of peoples when the neighbouring countries had long since formed larger nation states.

HISTORY

Maps are also a wonderful tool for exploring the past. The geography of historical maps is still recognisable to us because in geological terms not much has changed in the contours of the continents in the past 10,000 years. We therefore encounter familiar shapes and sizes in historical maps. When the geography is a bit off that's probably because mapping the world before GPS wasn't as precise. Besides the occasional geographic error, only the names of the countries, states or kingdoms that stretch across the continents are different. On the following maps you may find your home and favourite vacation spots, but they are put into a dramatically different context — be it as a place where the Vikings once attacked, as victims of the plague or as mysterious unknown place as of 1660.

Throughout the course of history, borders have been shifted and redrawn again and again; victors and conquerors have founded new countries, mostly from the remnants of the conquered areas. Maps showing this can help us better understand the present because they show us where the old borders were, and where they still have an impact today.

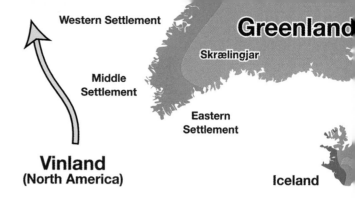

ATLANTIC
OCEAN

Scandinavian Settlement

- ▮ eighth century
- ▮ ninth century
- ▮ tenth century
- ▢ eleventh century
- ▮ denotes areas subjected to frequent Viking raids but with little or no Scandinavian settlement

The Viking Expansion

The so called 'Viking expansion' was led by Norse explorers, traders and warriors. Commonly lumped together as Vikings, they sailed most of the North Atlantic, travelled through part of the Mediterranean, even reached Russia. Vikings made it as far as the Caspian and Black Sea. The Vikings acted as looters, traders, colonists and mercenaries. In whatever function they

NORWEGIAN
SEA

Saami (Lapps)

Permia (Bjarmland)

Faroes

Shetland

Orkneys

Finns

Norway

Sweden

Chuds

Volga
Bulgars

Rus'
States

NORTH
SEA

Skane

Denmark

Letts

Lithuanians

Khazar
Khaganate

England

Prus

CASPIAN
SEA

Wends

Frisia

East Slavs

Normandy

West Slavs

Francia

Bulgars

Shirvia

South Slavs

BLACK SEA

Daylam

Italy

Byzantine Empire

Africa

Sicily

Abbasid
Caliphate

MEDITERRANEAN SEA

travelled, their navigational achievements remain somewhat underappreciated. A Viking whose name should be known more widely was Bjarni Herjólfsson. He was the first known European discoverer of the mainland of the Americas, which he sighted in 986. That means the Vikings spotted America at least 506 years before Christopher Columbus. In recent years countless wonderful popular history books on Vikings were published. Endless reading material for those who want to dive deeper into Viking history.

Columbus' Journeys

In 1492, Italian explorer Christopher Columbus convinced the Spanish court to finance his transatlantic maritime expeditions. While aiming to find a cost-saving shortcut to India, Columbus accidentally stumbled across the Americas, claiming the New World for his financiers. Columbus didn't set foot on the American mainland, in today's Venezuela, until his third voyage in 1498, and he never actually entered the present-day United States. He died in 1506, soon after his fourth trip, and it was only then, after the Americas had been named after Amerigo Vespucci, that Europeans realised the New World was a continuous landmass. Columbus' original goal, to find a westward route to Asia wasn't completed until 1521. The Castilian Magellan-Elcano expedition achieved even more as it managed to complete the first circumnavigation of the world.

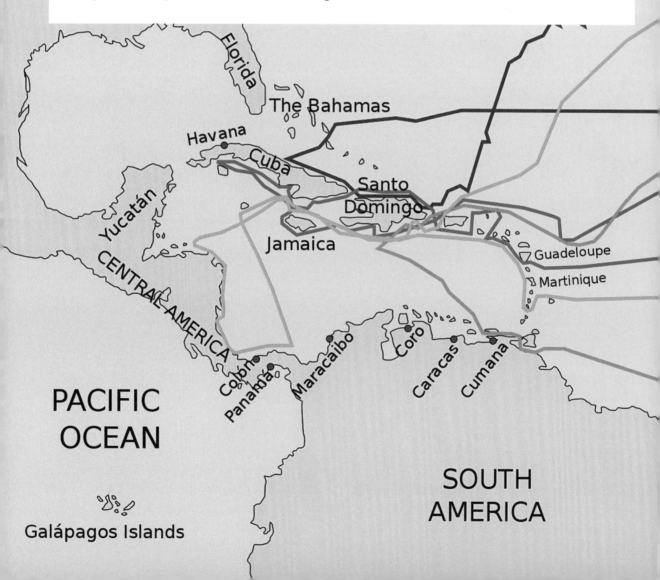

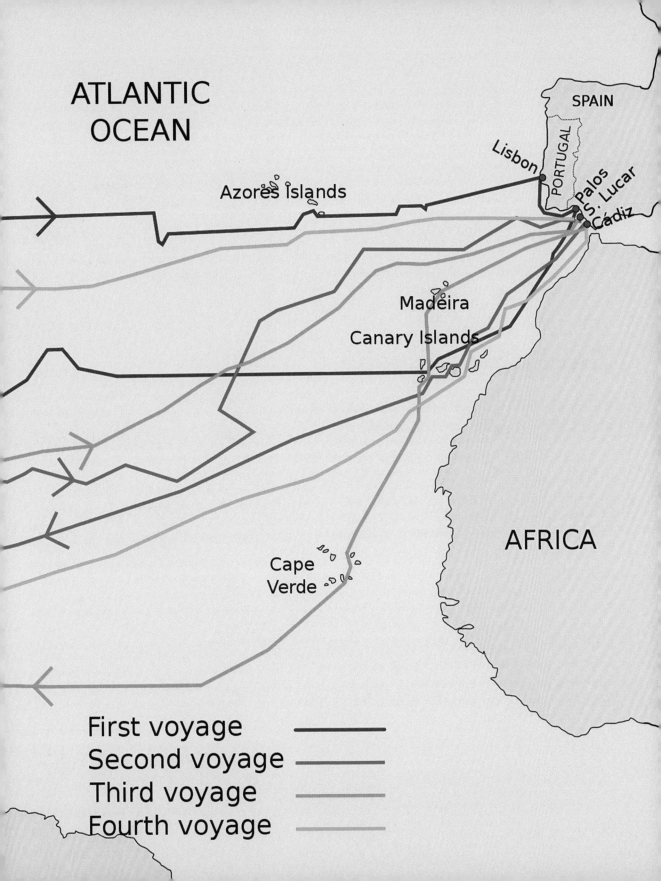

The Plague in Europe

In the 14th century, Europe fell victim to the largest pandemic in history (to date). The bubonic plague swept through entire countries, killing an estimated 25 to 50 million people. The higher end of these estimates would correspond to 60 per cent of the entire population. The bacterium responsible for the epidemic, Yersinia pestis, originated in Central Asia and came to Europe via trade routes such as the Silk Road. The map shows the spread of the pathogen and thus also the medieval trade routes. In the cities, the so-called Black Death was particularly devastating. This was where the risk of infection was greatest, while people back then had no knowledge about the spread or control of epidemics.

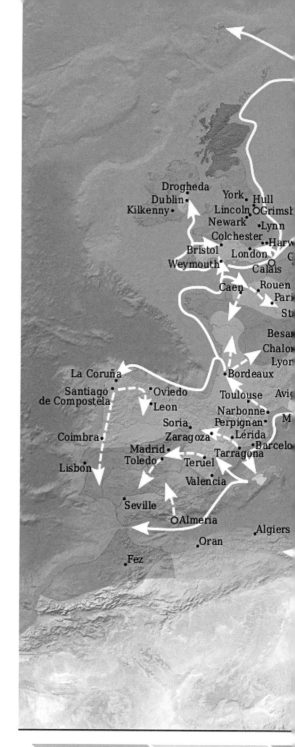

1346 **1347**

approximate border betwe
Principality of Kiev and th
passage forbidden for Chr

Trondheim
Hamar
Oslo
Göteborg
Halmstad
Lund
København
Stralsund Gdansk
Lubeck
Toruń
Frankfurt
Erfurt
Wroclaw
Prague
Kraków
Nürnberg
Vienna
Munich
Mürztal
Budapest
Oradea
Venice
Florence Split
Pisa
Perugia
Dubrovnik
Rome
Naples
Messina
Tunis
Tripoli

Stockholm
Visby
Kaliningrad
Elblag
Warsaw
Belgrade

Belozersk
Velikiy Novgorod   Nizhny Novgorod   Kazan
Pskov   Tver   Suzdal
Moscow   Ryazan
Smolensk
Hloukhiv
Chernihiv
Kiev
New Sarai
Old Sarai
Azov   Astrakhan

No data

Bucharest   Feodosia
Istanbul   Trabzon

Aleppo   Baghdad
Homs
Damascus
Jerusalem
Gaza
Al Marj   Alexandria
Cairo
Asyut

48   1349   1350   1351   1352   1353

 Land trade route

Trade route by water

n Horde –

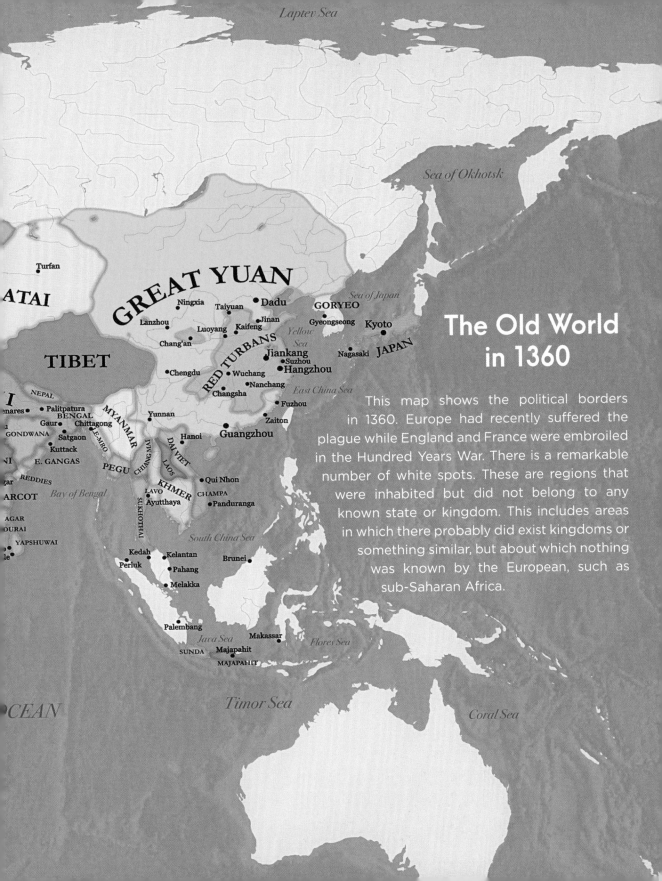

The Old World in 1360

This map shows the political borders in 1360. Europe had recently suffered the plague while England and France were embroiled in the Hundred Years War. There is a remarkable number of white spots. These are regions that were inhabited but did not belong to any known state or kingdom. This includes areas in which there probably did exist kingdoms or something similar, but about which nothing was known by the European, such as sub-Saharan Africa.

Laptev Sea

Sea of Okhotsk

Turfan

ATAI

GREAT YUAN

Ningxia • Taiyuan • Dadu

Sea of Japan

GORYEO

Lanzhou

Jinan

Gyeongseong • Kyoto

Yellow Sea

Luoyang • Kaifeng

Chang'an

Nagasaki • JAPAN

TIBET

Chengdu • Wuchang

Jiankang
Suzhou
Hangzhou

RED TURBANS

Nanchang

Changsha

East China Sea

NEPAL

Benares • Palitpatura

BENGAL

Gaur • Chittagong

GONDWANA

Satgaon

Kuttack

NI • E. GANGAS

REDDIES

ARCOT

AGAR
OURAI

YAPSHUWAI

Yunnan

Fuzhou

Zaiton

MYANMAR

Hanoi • Guangzhou

TEMRO

DAI VIET

PEGU

LAOS

KHMER

Qui Nhon

CHAMPA

LAVO • Panduranga

SUKHOTHAI

Ayutthaya

Bay of Bengal

South China Sea

Kedah • Kelantan • Brunei

Perluk • Pahang

Melakka

Palembang

Java Sea • Makassar

SUNDA • Majapahit

MAJAPAHIT

Flores Sea

OCEAN

Timor Sea

Coral Sea

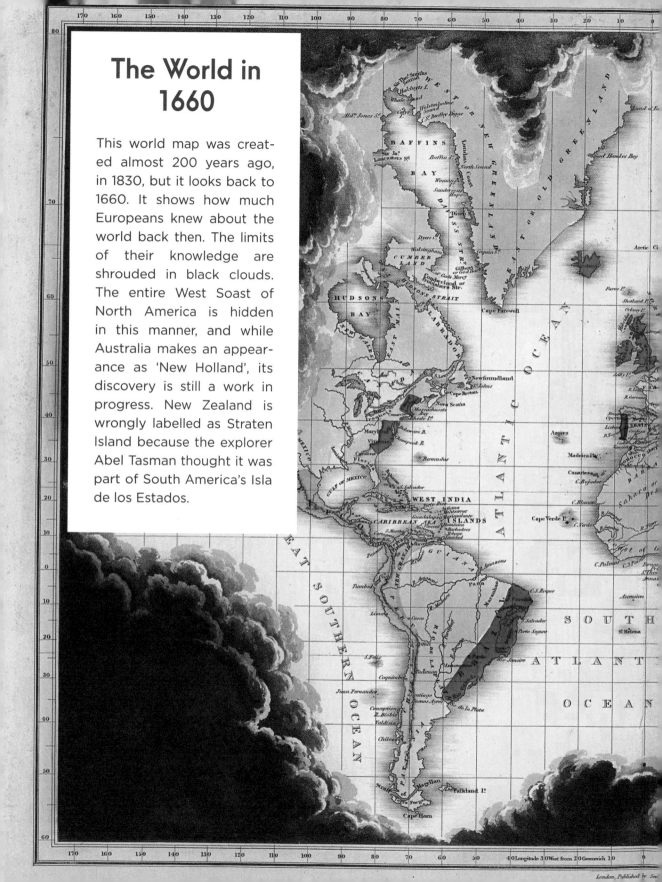

The World in 1660

This world map was created almost 200 years ago, in 1830, but it looks back to 1660. It shows how much Europeans knew about the world back then. The limits of their knowledge are shrouded in black clouds. The entire West Soast of North America is hidden in this manner, and while Australia makes an appearance as 'New Holland', its discovery is still a work in progress. New Zealand is wrongly labelled as Straten Island because the explorer Abel Tasman thought it was part of South America's Isla de los Estados.

London, Published by Son.

Engraved by Sid. Hall, Bury St. Bloomsb.

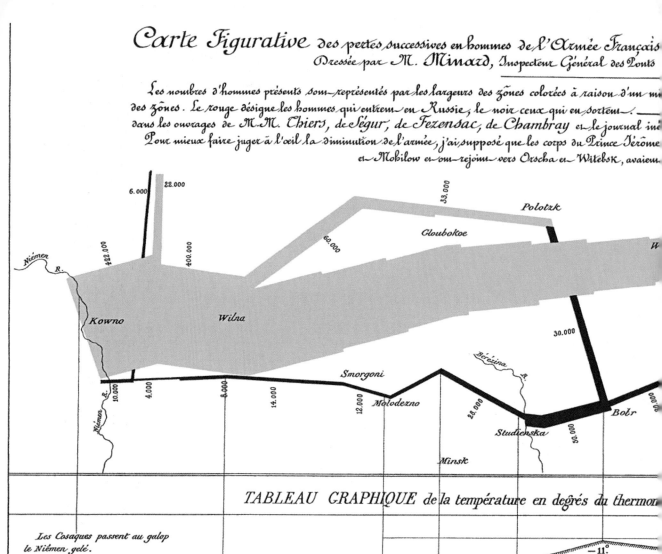

Napoleon's Russian Campaign

One of the best and earliest infographics is this map by the French civil engineer Charles Joseph Minard (1781—1870). His map explains Napoleon's Russian campaign from 1812 to 1813 and how it turned into an absolute catastrophe. This single graph includes many different variables. The beige line shows the dwindling number of troops as the French army marched

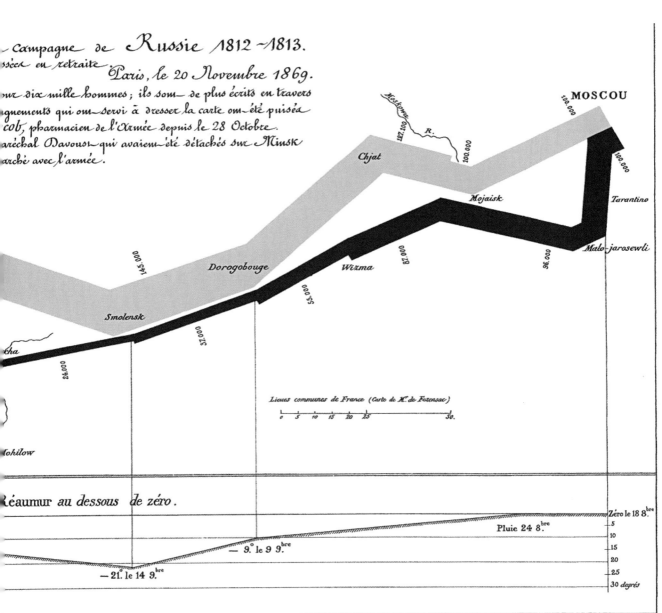

Imp. Lith. Regnier et Dourdet.

to Russia; the black line shows the troop strength on the retreat. Separation and reunification of units are clearly visible. At the bottom we see the altitude and, most importantly, the temperature, which is specified in Réaumur — the low point of minus 30 degrees Réaumur corresponds to minus 37.5 degrees Celsius. This is one reason why the retreat became so devastating for the French.

North America – if the Allied Powers Had Lost WWI (1916)

On February 10, 1916, *Life* magazine scared its readers by printing this fictional map on its cover. It served as a propaganda call to the USA to overcome its isolationism and enter World War One on the side of the Triple Entente (an alliance between England, France and Russia). The map presents a terrifying vision of what might happen if the German Empire, Austria-Hungary, Bulgaria and the Ottoman Empire won the war. The USA would be renamed New Prussia and every American city would meet the same fate; even the Atlantic would not be spared the rebranding. The other Central Powers also get pieces of North America, albeit much smaller ones. The state of New Mexico becomes a reservation for Americans. Why Canada would become a barbaric wilderness, however, is unclear; nor does it make much sense that the West Coast is being taken over by Japan, since Japan had been fighting on the side of the Entente powers since 1914.

North Korea
South Korea
Military Aid (North Korea)
Medical Aid (North Korea)
Other Aid (North Korea)
Military Aid (South Korea)
Medical Aid (South Korea)
Other Aid (South Korea)
Medical Aid (South Korea) and Other Aid (North Korea)

Countries Involved in the Korean War

In 1950, the Korean War began on the Korean Peninsula. It was not an 'ordinary' civil war, but one of the bloody proxy wars of the Cold War. The Soviet Union and the People's Republic of China supported communist North Korea, while the United Nations and above all the USA backed democratic South Korea. The map shows how global this conflict became as

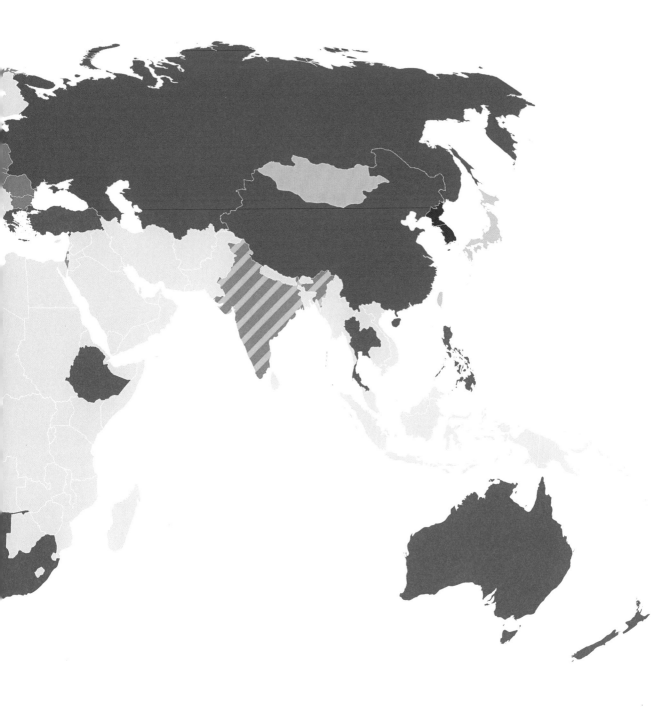

the world powers used their military strength to directly influence the war. The Korean War ended on July 27, 1953 with a ceasefire agreement. The peninsula remains divided, and the border is still closely guarded. The trench warfare came at a high cost: not only was almost all of Korea's industry destroyed, but about 940,000 soldiers and about three million civilians lost their lives.

The Last Executions in Europe

Today all but two European countries have abolished the death penalty. Russia has had a moratorium on executions since 1999, although there is no official law against it. Belarus is the sole outlier, as it continues to apply the death penalty. Jakub Marian compiled the dates of the most recent executions and turned them into a beautiful map. It tells us both the year and the method of the last execution in each country. In East Germany (1972 on the map), the last actual execution was carried out on an intelligence officer in 1981. In West Germany, the last person was executed in 1949 for murder. In France, the last execution took place in 1977 via Guillotine, 185 years after the first use of the machine during the French Revolution in 1792. The death penalty is becoming increasingly unpopular around the world. 106 countries have already abolished it, others no longer apply it. Around the world 56 countries continue to sentence people to death, with China being the country that executes the most people.

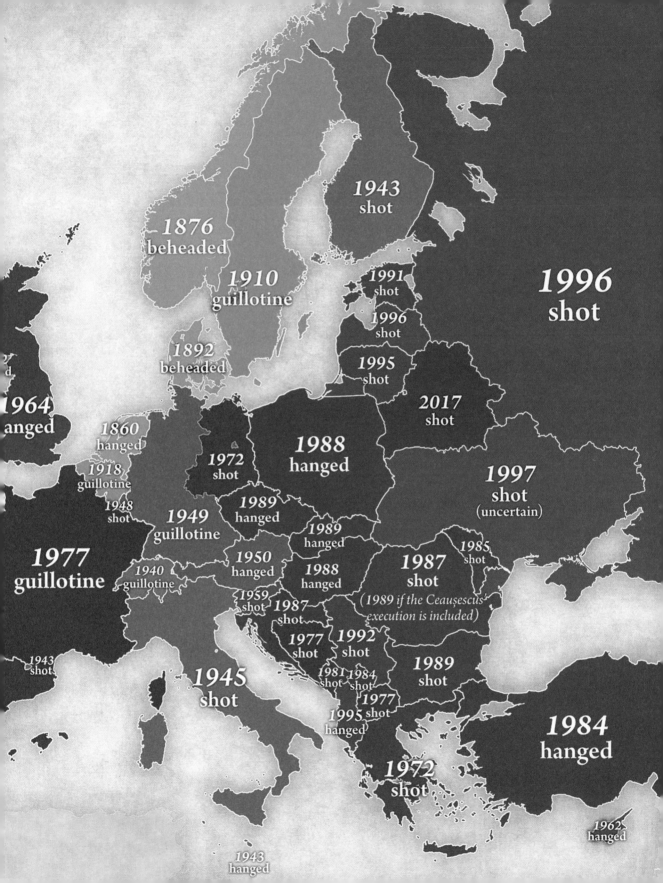

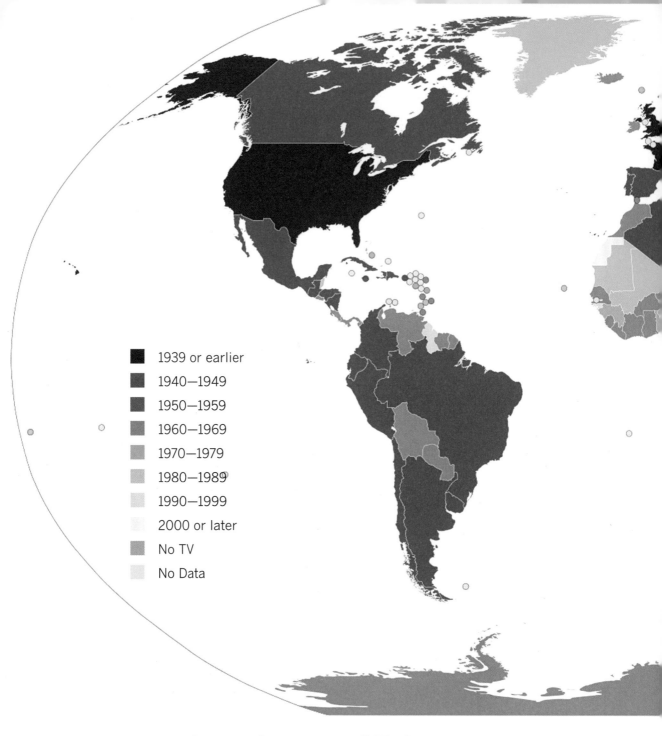

Introduction of Television

This map shows the decade in which countries got their first official, regular television channel. The first trial run began in the USA in 1928. Regular operation began in 1938, but then only in New York and Los Angeles. All major nations have been broadcasting regularly since the 1930s. Radio was the

Legend:

- 1939 or earlier
- 1940–1949
- 1950–1959
- 1960–1969
- 1970–1979
- 1980–1989
- 1990–1999
- 2000 or later
- No TV
- No Data

first non-written means of mass communication, and television added pictures. Any person on TV could be seen by the whole country. With the introduction of more and more channels and then the Internet, the appeal of television decreased. The newest country on the map is South Sudan, which recently declared its independence. In such a case, starting a national television station continues to be a powerful symbolic act.

Hannibal's Route of Invasion

This map shows one of the greatest military feats in history. After the Carthaginians fought the Romans in Spain, Hannibal's army in 218 BC made an unexpected move that put him forever into the history books. He marched his troops, including cavalry and most famously his African war elephants, across a high pass in the Alps to take on the Romans on their very own Italian peninsula. The Romans thought of the Alps as a secure natural fortress that protected them against invasion via the northern land route. Nobody expected Hannibal's bold (shall we say mad?) move. In December he defeated Roman armies in the north. Most of his famous war elephants died of cold or disease during the winter while Hannibal continued his campaign through Italy. For 15 years he wreaked havoc, allegedly killing or wounding over a million Roman citizens but without taking Rome itself. The second Punic war ultimately ended with Rome as the victor.

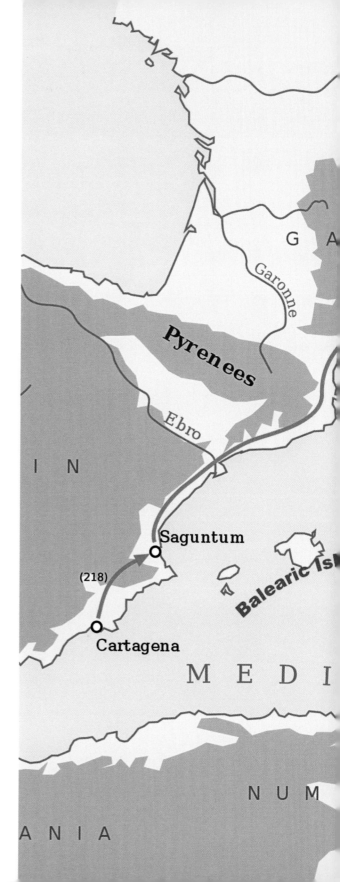

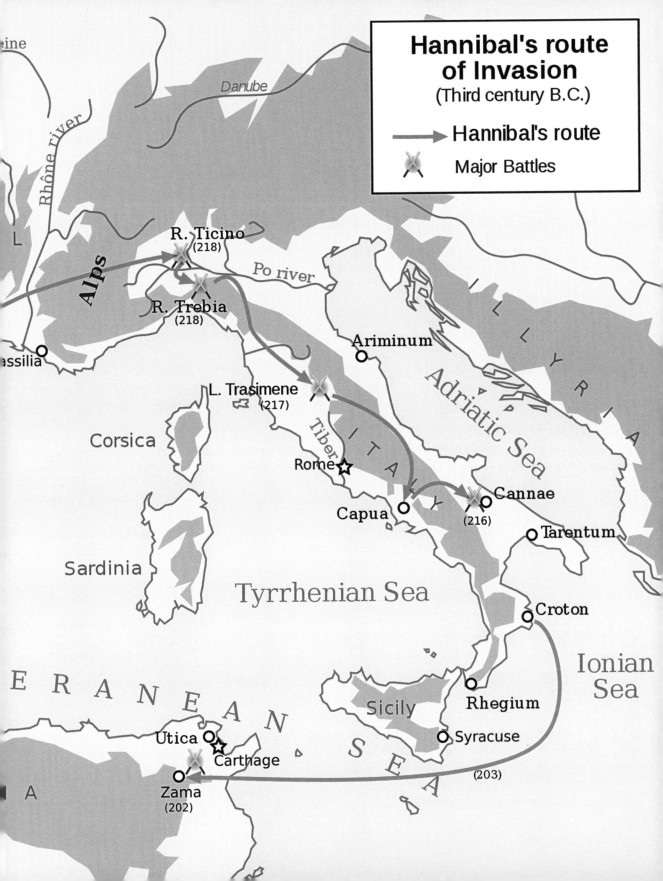

Hannibal's route of Invasion
(Third century B.C.)

→ Hannibal's route
⚔ Major Battles

Rhône river
Rhine
Danube
Alps
R. Ticino (218)
Po river
R. Trebia (218)
Massilia
Ariminum
L. Trasimene (217)
Corsica
Tiber
Rome
ITALY
Capua
Cannae (216)
Tarentum
ILLYRIA
Adriatic Sea
Sardinia
Tyrrhenian Sea
Croton
Ionian Sea
Sicily
Rhegium
Syracuse
(203)
MEDITERRANEAN SEA
Utica
Carthage
Zama (202)

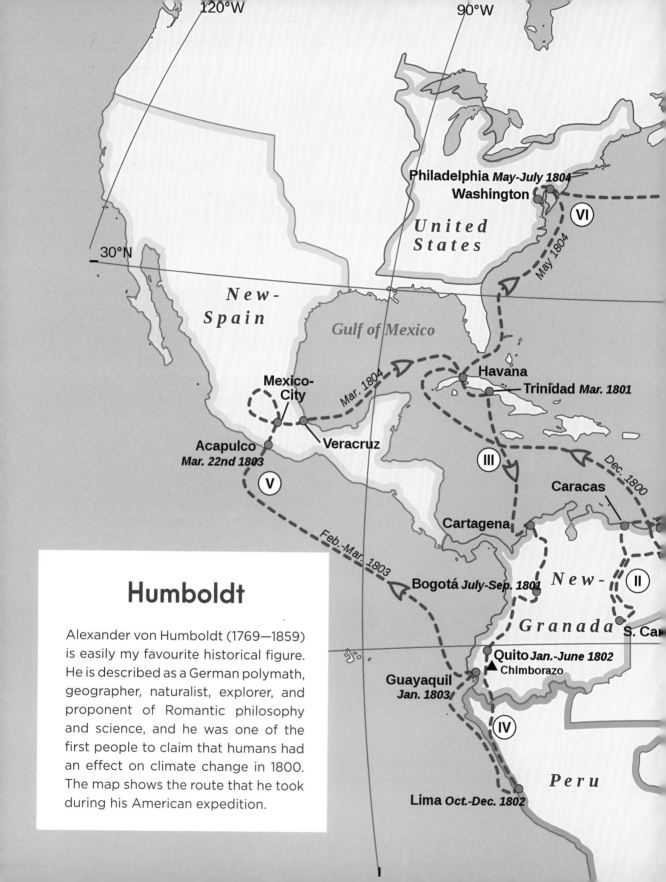

Humboldt

Alexander von Humboldt (1769—1859) is easily my favourite historical figure. He is described as a German polymath, geographer, naturalist, explorer, and proponent of Romantic philosophy and science, and he was one of the first people to claim that humans had an effect on climate change in 1800. The map shows the route that he took during his American expedition.

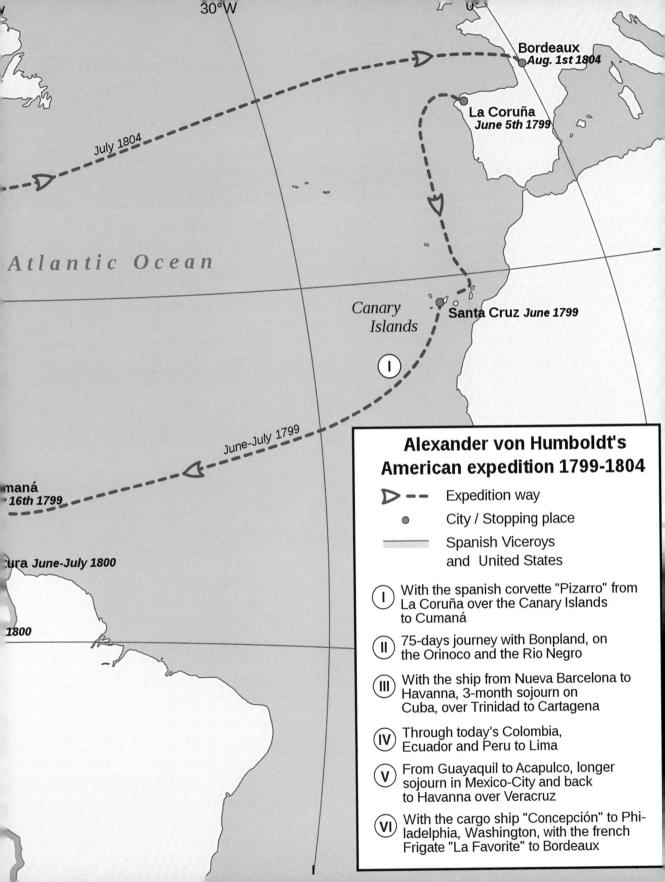

Alexander von Humboldt's American expedition 1799-1804

▷ - - Expedition way

● City / Stopping place

▬ Spanish Viceroys and United States

ⓘ With the spanish corvette "Pizarro" from La Coruña over the Canary Islands to Cumaná

ⓘⓘ 75-days journey with Bonpland, on the Orinoco and the Rio Negro

ⓘⓘⓘ With the ship from Nueva Barcelona to Havanna, 3-month sojourn on Cuba, over Trinidad to Cartagena

ⓘⓥ Through today's Colombia, Ecuador and Peru to Lima

Ⓥ From Guayaquil to Acapulco, longer sojourn in Mexico-City and back to Havanna over Veracruz

Ⓥⓘ With the cargo ship "Concepción" to Philadelphia, Washington, with the french Frigate "La Favorite" to Bordeaux

30°W

Bordeaux
Aug. 1st 1804

La Coruña
June 5th 1799

July 1804

Atlantic Ocean

Canary Islands

Santa Cruz *June 1799*

June-July 1799

ⓘ

...maná
16th 1799

...ura *June-July 1800*

1800

AFTERWORD

For now, we have reached the end of our cartographic journey together. It spanned over 100 maps — from detailed world maps based on complex data to simple schematic maps just meant to entertain you. In the introduction, I mentioned that I hope to take away the fear of data from those who are terrified of numbers by presenting data in the shape of interesting maps. I would be delighted to hear from you whether I succeeded. Just send me a quick message by email or on Twitter.

I also hope that some maps were more than just entertainment and that they allowed you to explore well-known topics from a new perspective. Perhaps this compilation even rekindled your love of maps.

There have probably been a few maps where you questioned the accuracy of the data. I would like all my readers to share this healthy, sceptical attitude. We should always look critically at data sources and maps. It's worth asking: is this meant to manipulate me? At the same time, we should avoid the temptation to dismiss any data that does not correspond with our worldview as fake news or just pure nonsense. It is not always easy for me to accept data and maps that force me to reconsider my opinion. But I do always regard it as an invitation to read up on a new topic and expand my worldview with a few new facts and perspectives. After all, even the best map is only the first step into a new adventure, into a new subject area. As I have already mentioned in the introduction: we must not confuse the map with the territory. No map tells us the whole story, no map truly represents reality. We still must think for ourselves to really understand what the map is trying to tell us.

If you want even more maps: I post half a dozen new fascinating maps and data animations on my Twitter and Facebook accounts every day. I'd be delighted to see you in the digital realm.

 @SimonGerman600: As many maps and data as your screen can hold

 Simon shows you maps: As on Twitter, maps galore

 Simon Kuestenmacher: This is my professional account where I write about global and Australian demographics — feel free to connect

 Simon Kuestenmacher: How exciting is the life of a demographer and map junkie? Judge for yourself

ACKNOWLEDGEMENTS

I would like to extend heartfelt thanks to all the cartographers who were so kind as to make their works available for this book. I spend several hours every day studying and admiring the works of various cartographers and other data aficionados and sharing them with fellow map enthusiasts on Twitter and Facebook. In this book, I have only presented a small selection. Since this is a printed volume, I could of course not include any of the countless animated maps that I come across.

SOURCES

P. 12/13: 'Population Density': NASA Earth Observations (NEO)/Socioeconomic Data and Applications Center (SEDAC) / CIESIN Columbia University, December 2018

P. 14: 'Three Continents': Bibi Saint-Pol/Wikipedia; 'Four Continents' and 'Five Continents': shutterstock.com/Peter Hermes Furian

P. 15—19: all maps: shutterstock.com/Peter Hermes Furian

P. 20/21: 'Black Marble': NASA Earth Observatory

P. 22/23: 'More People Live Inside the Circle Than Outside of It': shutterstock.com/ Volodymyr Nikulishyn

P. 26/27: 'Arable Land': www.croplands.org, Thenkabail, P. S., Teluguntla, P., Oliphant, A., Xiong, J., 2020. Landsat Satellite derived Global Cropland Extent Product at 30 m developed using Machine Learning Algorithms and Cloud computing

P. 29: 'Global Shipping Routes until 1860' und 'Global Shipping Routes from 1945 to 2000': Benjamin Schmidt, 20 Cooper Square, New York University, New York

P. 30/31: 'Earthquakes': This map is intellectual property of Esri. Copyright © 2019 Esri and its licensors. All rights reserved. John Nelson (@John_M_Nelson): AdventuresInMapping. com, Data: USGS (earthquake.USGS.gov/earthquakes), Image: NASA Visible Earth (VisibleEarth.nasa.gov)

P. 32/33: 'Telegraph Cables and Steamer Routes in 1924': Rand McNally & Co.'s *Commercial Atlas Of America*. Rand McNally Company, Chicago, 1924 (www.davidrumsey.com)

P. 34/35: 'Submarine Internet Cables': TeleGeography (www.telegeography.com)

P. 36/37: 'Global Air Traffic': Grandjean Martin, *Connected World: Untangling the Air Traffic Network*, 2016, http://www.martingrandjean.ch/connected-world-air-traffic-network/

P. 38/39: 'Global Consequences of Climate Change': Parag Khanna, author of *Connectography: Mapping the Future of Global Civilization* (2016)

P. 40/41: 'Solar Power': Photovoltaic power potential © 2019 Solargis

P. 44/45: 'Passport Power': Map: shutterstock.com/Andrei Minsk; Data: Henley Passport Index 2019 Data

P. 46/47: 'Tipping': Soshial/original work; Data: Wikivoyage Tipping sections

P. 48/49: 'World Socket Map': shutterstock.com/sunsinger

P. 50/51: 'International Country Codes': Chumwa/Wikipedia (CC BY-SA 2.0)

P. 52/53: 'Left- and Right-Hand Traffic': Benjamin D. Esham (bdesham); Data: Sens de circulation.png

P. 54/55: 'Literal Meaning of European Country Names': NeoMam Studios production for Credit Card Compare (www.creditcardcompare.com.au)

P. 56/57: 'Hippie Trail': NordNordWest/Wikipedia (CC BY-SA 3.0)

P. 58/59: 'Average Waiting Time for Hitchhikers': Ábel Sulyok (Ungarn, 2019); hitchwiki.org

P. 61: 'European Long-Distance Cycling Routes': EuroVelo, the European cycle route network, www.eurovelo.org, 2019

P. 62/63: 'Speed Limits Worldwide': Map: shutterstock.com/ArtMari; Data: Terrorist96/ Wikipedia

P. 64/65: 'Travel Times from London in 1914': John George Bartholomew (1860—1920), in: *An atlas of economic geography, London*: Oxford University Press, 1914

P. 66/67: 'Travel Times from London in 2016': Rome2rio.com, 2016

P. 68/69 'The Voyage of the Beagle': Benjamin Schmidt, 20 Cooper Square, New York University, New York

P. 71: 'Eruption of the Yellowstone Volcano': Mastin et al., 2014 Yellowstone paper in: Mastin, L. G., Van Eaton, A. R. and Lowenstern, J. B., 2014. Modelling ash fall distribution from a Yellowstone supereruption. Geochemistry, Geophysics, geosystems, 15(8): 3459—3475

P. 74/75: 'Arctic Migratory Birds': Conservation of Arctic Flora and Fauna (CAFF), 2013. Arctic Biodiversity Assessment: Contribution for: Policy Makerp. CAFF, Akureyri, Island

P. 76/77: 'Wolf Pack Territories': Voyageurs Wolf Project

P. 78/79: 'Countries With Native Species of Wild Cats': Robert Parkin

P. 80/81: 'Flamingo World Map': Phoenix_B_1of3 (talk)/original work; Data: Fred Cooke, Jenni Bruce: The Encyclopedia of Animals: a complete visual guide

P. 82/83: 'Pleistocene Mammoth Habitats': Azcolvin429/Wikipedia (CC-BY-SA 4.0)

P. 84/85: 'Lifestock Density': Gilbert, M., Nicolas, G., Cinardi, G., Van Boeckel, T. P., Vanwambeke, S. O., Wint, G. R. W., Robinson, T. P., 2018. Global distribution data for cattle, buffaloes, horses, sheep, goats, pigs, chickens and ducks in 2010. Scientific Data 5, 180227. https://doi.org/10.1038/sdata.2018.227

P. 86/87: 'The World from the Point of View of a Dolphin': Frans Blok/3 Develop, www.3develop.nl

P. 88/89: 'World Animals': © Kentaro Nagai/Graflex Directions

P. 90/91: 'Number of Venomous Species Per Country': Reddit/lanson15

P. 92: 'The Extermination of the American Bison' based on image 'Bisons.JPG', which was based on William Temple Hornaday's book: Citynoise (talk)/English Wikipedia (CC-BY-SA-3.0)

P. 93: 'The Australian Dingo Fence': Travel Encyclopedia; edited for translation

P. 97: 'Pangaea With Modern Borders': Massimo Pietrobon

P. 99: 'Antipodes World Map': AnonMoos/Wikipedia (CC BY-SA 2.5)

P. 100/101: 'Date Line': Map: Google Earth

P. 103: 'Canada Lives Below the Red Line': Map: shutterstock. com/Abdulloh Jehyapar

P. 105: 'Greenland is Farther East, West, North and South than Iceland': Map: shutterstock.com/Cvijovic Zarko

P. 107: 'South America is also East America': shutterstock.com/dikobraziy

P. 108/109: 'Ten Bands of Latitude with Equal Land Area': Neil Kaye (@neilkaye)

P 110/111: 'Australian Vegetation': The New Oxford wall maps of Australia [cartographic material]. Vegetation (1929)/Griffith Taylor

P. 112: 'Heihe Tengchong Line': Ginbayashi/Wikipedia

P. 114/115: 'Mercator Projection': Neil Kaye (@neilkaye)

P. 117: 'The True Size of Africa': Czech is Cyrillized/Wikipedia (CC BY-SA 1.0)

P. 119 'Rivers and Lakes of the World': James Reynolds/John Emslie; Publisher: James Reynolds; Source: David Rumsey Map Collection

P. 120/121: 'Football With Neil and Buzz': NASA/ALSJ

P. 122/123: 'Overlay of the Roman Empire and the US': Massimo Pietrobon

P. 124/125: 'Great Pacific Garbage Patch': Scientific Reports 8, Article number 4666 (2018), L. Lebreton, B. Slat, F. Ferrari, B. Sainte-Rose, J. Aitken, R. Marthouse, S. Hajbane, S. Cuncolo, A. Schwarz, A. Levivier, K. Noble, P. Debeljak, H. Maral, R. Schoeneich-Argent, R. Brambini, J. Reisser (CC BY-SA 4.0)

P. 128/129: 'Coffee and Wine Belts' Map: shutterstock.com/Peter Hermes Furian

P. 130/131: 'Tongue-in-Cheek Alcohol Preferences of Europe': shutterstock.com/g/milosmilovanovic

P. 133: 'Actual Alcohol Preferences of Europe': Reddit: @NaytaData

P. 134/135: 'Obesity in Europe': Reddit: John Leith

P. 138/139: 'Currencies Before the Euro': Map: shutterstock.com/g/milos+milovanovic, Data: Wikipedia

P. 140/141: 'Difficulty of European Languages': Map: shutterstock.com/g/milos+milovanovic, Data: Wikipedia

P. 143: 'Blue Banana': Map: shutterstock.com/g/milos+milovanovic, Data: Wikipedia

P. 144/145: 'What Would Other Brexits Be Called?': Map: shutterstock.com/g/milos+milovanovic

P. 146/147: 'Number of Letters in European Alphabets': Map: shutterstock.com/g/milos+milovanovic

P. 148/149: 'The Surname "Smith" in All European Languages': Map: shutterstock.com/g/milos+milovanovic, Data: Wikipedia

P. 150/151: 'Fantasy Map of Europe': Tautvydas Norkus

P. 152/153: 'French Kissing': Bill Rankin, www.radicalcartography.net

P. 154/155: 'Who Brings the Europeans Their Christmas Presents?': Jakub Marian, jakubmarian.com

P. 156/157: 'What is Germany called in Other European Languages?': Shyamal/Wikipedia (CC BY-SA 3.0)

P. 160/161: 'The Viking Expansion': Phirosiberia/Wikipedia, 8 December 2009, 13:12 (UTC)

P. 162/163: 'Columbus' Journeys': Viajes de colon en.svg/Wikipedia (CC BY-SA 3.0)

P. 164/165: 'The Plague in Europe': Flappiefh/Wikipedia (CC BY-SA 4.0)

P. 166/167: 'The Old World in 1360': Jan Wignall

P. 168/169: 'The World in 1660': Quin, Edward: A.D. 1660. At The Restoration Of The Stuarts. Engraved by Sidy. Hall Bury Strt. Bloomsby. London, Published by Seeley & Burnside, 169, Fleet Street, 1830. ; Source: David Rumsey Historical Map Collection

P. 170/171: 'Napoleon's Russian Campaign': Charles Joseph Minard 1869

P. 172/173: 'North America if the Allied Powers Had Lost WWI (1916)': Cornell University — PJMode Collection of Persuasive Cartography

P. 174/175: 'Countries Involved in the Korean War': Snehasish Chowdhury

P. 176/177: 'The Last Executions of Europe': Jakub Marian, jakubmarian.com

P. 178/179: 'Introduction of Television': Laydan Mortensen/Wikipedia

P. 180/181: 'Hannibal's Route of Invasion': Wikipedia/Abalg + traduction made by Pinpin

P. 182/183: 'Humboldt': Alexander von Humboldt expedition-en.svg/Wikipedia (CC-BY-SA-2.5)/Alexrk translated by Cäsium137

This English language edition Published in 2021 by OH Editions,
an imprint of Welbeck Non-Fiction Limited,
part of Welbeck Publishing Group
20 Mortimer Street
London W1T 3JW
English Translation by © Welbeck Non-Fiction Limited

First published by © Riva Verlag in 2020
Original title: *Mad Maps*
www.m-vg.de

Random Maps
ISBN 978-1-91431-706-4
Text © Simon Kuestenmacher
Translator: Nadja Rehberger
Design: Stuart Hardie
Production: Rachel Burgess

A CIP catalogue record for this book is available from the British Library

Printed in Dubai

10 9 8 7 6 5 4 3 2 1